Katja Yafimava

Post-Soviet
Russian-Belarussian Relationships

The Role of Gas Transit Pipelines

With a foreword by Jonathan P. Stern

SOVIET AND POST-SOVIET POLITICS AND SOCIETY

ISSN 1614-3515

Recent volumes

34   *Florian Mühlfried*
      Postsowjetische Feiern
      Das Georgische Bankett im Wandel
      Mit einem Vorwort von Kevin Tuite
      ISBN 3-89821-601-2

35   *Roger Griffin, Werner Loh, Andreas Umland (Eds.)*
      Fascism Past and Present, West and East
      An International Debate on Concepts and Cases in the Comparative Study of the Extreme Right
      With an afterword by Walter Laqueur
      ISBN 3-89821-674-8

36   *Sebastian Schlegel*
      Der „Weiße Archipel"
      Sowjetische Atomstädte 1945-1991
      Mit einem Geleitwort von Thomas Bohn
      ISBN 3-89821-679-9

37   *Vyacheslav Likhachev*
      Political Anti-Semitism in Post-Soviet Russia
      Actors and Ideas in 1991-2003
      Edited and translated from Russian by Eugene Veklerov
      ISBN 3-89821-529-6

38   *Josette Baer (Ed.)*
      Preparing Liberty in Central Europe
      Political Texts from the Spring of Nations 1848 to the Spring of Prague 1968
      With a foreword by Zdeněk V. David
      ISBN 3-89821-546-6

39   *Михаил Лукьянов*
      Российский консерватизм и реформа, 1907-1914
      С предисловием Марка Д. Стейнберга
      ISBN 3-89821-503-2

40   *Nicola Melloni*
      Market Without Economy
      The 1998 Russian Financial Crisis
      With a foreword by Eiji Furukawa
      ISBN 3-89821-407-9

41   *Dmitrij Chmelnizki*
      Die Architektur Stalins
      Bd. 1: Studien zu Ideologie und Stil
      Bd. 2: Bilddokumentation
      Mit einem Vorwort von Bruno Flierl
      ISBN 3-89821-515-6

Katja Yafimava

POST-SOVIET
RUSSIAN-BELARUSSIAN RELATIONSHIPS

The Role of Gas Transit Pipelines

With a foreword by Jonathan P. Stern

ibidem-Verlag
Stuttgart

Bibliografische Information Der Deutschen Bibliothek

Die Deutsche Bibliothek verzeichnet diese Publikation in der Deutschen Nationalbibliografie; detaillierte bibliografische Daten sind im Internet über <http://dnb.ddb.de> abrufbar.

Frontcover: Picture printed with kind permission from © Chatham House.

∞

Gedruckt auf alterungsbeständigem, säurefreien Papier
Printed on acid-free paper

ISSN: 1614-3515
ISBN-10: 3-89821-655-1
ISBN-13: 978-3-89821-655-5
© *ibidem*-Verlag
Stuttgart 2007
Alle Rechte vorbehalten

Printed in Germany

For my parents with love and gratitude

Contents

List of Tables and Maps   9
Acknowledgements   11
Foreword by Jonathan P.Stern   13

Introduction   15

1   **The Russian-Belarussian Relationship: Geopolitics
     and Geo-economics**   25

     1.1   The Break-up of the USSR and the Creation of
            the CIS: Perceptions and Interests   25

     1.2   The Russian-Belarussian Integration:
            Three-dimensional Evolution   29

2   **Belarus and 'Cheap' Russian Gas: Can Belarus
     Afford Not to Reform?**   43

     2.1   Analysis of the Belarussian Economic Situation   47

     2.2   Can Belarus Afford to Pay Increased Prices for
            Gas Imports?   51

     2.3   Implications of Gas Price Increases for the
            Sustainability of the Belarussian Economy   56

3   **Russian Gas to Europe: An Analysis of Existing
     and Projected Export Routes**   63

     3.1   Russia and the Gas Supply of Europe: Dependence,
            Diversification and Export Routes   64

     3.2   The Gas Transit Routes: Belarus and Ukraine   70

     3.3   The Offshore Export Route: Is the North European Gas
            Pipeline an Alternative to the Belarussian Transit Routes?   81

4   **How Significant Is Belarus's Gas Bargaining Power?**   89

     4.1   The Belarus-Gazprom Gas Dispute: Its History
            and Current State   89

     4.2   Factors that Push Belarus and Gazprom to Re-negotiate   99

     4.3   Belarus and the Role of the Russian Independent
            Gas Companies   103

     Conclusion   111

     Bibliography   119
     Appendix   131

List of Tables and Maps

Tables

2.1   EBRD Transitions Indicators: Belarus and
       Russia in Comparison   131

2.2   Dynamics of Gas Prices, Transit Fees and Volumes of Gas
       Supplied to Belarus by Gazprom and the Independent
       Russian Gas Companies   58

2.3   Dynamics of Gas Imports Expenditures   59

3.1   Russia's Gas Exports to Europe in 2004 and Their
       Share in Domestic Consumption (top 10 importers)   66

4.1   Dynamics of Exports of Russian Gas by Gazprom and
       Itera to Belarus   104

Maps

3.1   Russia's Gas Export Routes   68

3.2   The Ukrainian Gas Transit Network   71

3.3   The North European Gas Pipeline   82

Acknowledgements

I am greatly indebted for invaluable research advice and support to my supervisors, Professor Jonathan P. Stern, director of the natural gas programme at the Oxford Institute for Energy Studies, and Dr. Carol Leonard of St. Antony's College Oxford. I also want to thank Prof. Robert Mabro and Dr. Robert Skinner for their kind permission to use the Institute's library. My research also benefited from the advice of Prof. Mario D. Nuti on the developments of the Belarussian economy. Finally, I owe special thanks to Dr. Richard Michaelis whose valuable suggestions made this book altogether more readable.

It is my pleasure to thank the Oxford University Press and the Universities UK Scheme who funded my studies at the University of Oxford with the Clarendon Fund Scholarship and the Overseas Students Award respectively.

Foreword

Until the early 2000s, Russian-Belarussian relations, including the natural gas relationship between the countries, have been issues to which the majority of Europeans – even those involved in natural gas trade – paid little attention. That situation began to change with the building of the Yamal-Europe pipeline and accelerated with the price and supply problems between the countries in early 2004. With the sharp deterioration in Russian-Ukrainian relations since 2005, the route for Russian natural gas supplies to Europe through Belarus has assumed a much more important status. At the same time, Russian impatience with, and distrust of, the behaviour of all CIS transit countries has led to a decision to move new European transportation infrastructure offshore, hence the decision to build (two) North European Gas Pipelines through the Baltic Sea to Germany.

In the mid-2000s, Belarus remains the only country, within a broadly defined European political, economic, and energy space, to have clearly stated that it is not seeking European Union or NATO membership; and (like the Russian Federation) the country has not yet ratified the Energy Charter Treaty. At the same time, among the post-Soviet European states, Belarus remains Russia's only unambiguously loyal political and strategic ally. This very clear position places Belarus firmly outside not just EU political frameworks, but also the EU legislative and regulatory *acquis* for the utility industries.

This book examines a country, and a bilateral relationship, whose strategic and logistical importance has barely been recognised by the majority of European Union countries on which it borders. It will be a valuable reference not just for the information it provides about Belarus and for those interested in post-Soviet relations between the two countries, but also for the more specific dynamics of bargaining between the two countries over gas and wider economic and political issues. The significance of these developments for natural gas trade is only in the mid-2000s starting to be recognised and acknowl-edged.

Katja Yafimava's book is the only detailed account in English of this very im-portant relationship for European gas and energy trade. She deserves con-siderable credit for her diligence and perseverance which have produced this work at such an important time for Russian exports, and European imports, of natural gas.

Jonathan Stern
May 2006
Oxford

Introduction

During the first post-USSR decade, the Russian gas company, Gazprom, which enjoys a monopoly on gas exports to Europe, saw Belarus, unlike a recalcitrant Ukraine, as a reliable transit route. As a result Gazprom undertook a substantial investment in a new gas pipeline system in Belarus, Yamal-Europe. This new route was seen as a major strategic export corridor that would serve Gazprom's growing exports to Europe thus enabling it to diversify its export routes and reduce dependence on unreliable Ukrainian ones. However, Belarus's reliability as a transit country came into question when, in February 2004, having failed to pay increased gas prices demanded by Gazprom, Belarus undertook unauthorised gas offtakes from Yamal-Europe. This led Gazprom to cut off all gas flows to Europe via Belarus. Even though gas blip was very brief and caused no serious disruption to European consumers, it threatened Gazprom's reputation as a reliable supplier and caused serious concerns in Europe regarding the reliability of Russian gas supplies.

The underlying cause of the conflict was Belarus's inability to pay for gas imports in full, resulting in the accumulation of significant debt. Up to 2004, in line with the 1995 Customs Union agreement, according to which all supplies (including energy) were to be traded in domestic prices of the manufacturing country, Gazprom supplied Belarus with gas at Russian domestic prices. This enabled Belarus to claim that Gazprom's refusal to supply gas to Belarus at domestic prices starting from 2004 violated the Customs Union agreement. However, there exists another, separate document, the Russian-Belarussian Intergovernmental Agreement, according to which Belarus was prepared to privatise its gas transit and transmission network, Beltransgaz, and create a 50/50 joint venture (JV) with Gazprom. Gas would continue to flow to Belarus at Russian domestic prices only if this were established. Curiously, the International Agreement does not contain any concrete information on the details of the JV creation, such as, for example, the method of asset evaluation, being rather an agreement *in principle*. Such uncertainty enabled Belarus to re-

nege on the agreement and to refuse to establish the JV. Indeed, Belarus motivated its refusal on the grounds that Gazprom's evaluation of Beltransgaz
was much lower than its market value. This, in turn, led Gazprom to stop shipping gas to Belarus at domestic prices and to require an initial export price increase.

Until the gas crisis the issue of the importance of Belarus as a gas transit
route, though attracted some scholarly attention[1], but certainly not as much as
it deserved. This is partly explained by the fact that Belarus, largely regarded
as the Russia's closest ally, was not perceived as a country that might act in a
manner that might cause Russia's dissatisfaction serious enough to cut off
gas supplies. Moreover, *Belarus as such* was not sufficiently explored in the
academic literature. This lack of interest is to some extent unsurprising given
the undemocratic developments in the country that the international community does not have enough leverage upon and its 'laggard' status in the transition to a market economy. However, the gas crisis of 2004 because it threatened the reliability of gas exports to several European countries vividly demonstrated Belarus's importance, underlying the necessity of bringing it back
into research focus and explaining how and why the seemingly amicable Russian-Belarussian relationship had deteriorated to such an extent as to cause
Russia to cut gas flows via Belarus and Belarus to label Russia's actions as 'a
terror act of highest order'.

The limited research that has so far been undertaken about Belarus deals
mostly with separate issues, such as Belarus's domestic politics[2], economics[3],

---

[1]   Margarita Balmaceda, 'Belarus as a Transit Route: Domestic and Foreign Policy
        Implications', in Margarita Balmaceda, J. Clem and L. Tarlow, eds., *Independent
        Belarus: Domestic Determinants, Regional Dynamics and Implications for the West*
        (Cambridge, MA: HURI/Davis Center for Russian Studies; Harvard University
        Press, 2002). pp. 162-196. See also Chloe Bruce, *Fraternal friction or fraternal fic-
        tion? The gas factor in Russian-Belarussian relations* (Vienna: University of Vi-
        enna, 2005).

[2]   Kathleen Mihalisko, 'Belarus: retreat to authoritarianism', in Karen Dawisha and
        Bruce Parrot, eds., *Democratic changes and authoritarian reactions in Russia,
        Ukraine, Belarus, and Moldova* (Cambridge: Cambridge University Press, 1997),
        pp. 223-281. See also Taras Kuzio, 'Belarus and Ukraine: democracy building in a

military and security issues[4]. It does not explore how these issues connect with each other or, indeed, how they influence gas trade and transit. However, since gas issues spill over into political, economic and military areas of Belarussian-Russian relations it is necessary to analyse them in all their linked complexity. Therefore our book employs an interdisciplinary approach that allows us to integrate these issues and analyse their effect on the reliability of Russian gas exports to Europe.

Belarus's bilateral relations with Russia have received somewhat more substantial scholarly attention[5], especially the Russian-Belarussian integration process. However, while conducting research on integration issues, scholars tended to underplay the issue of Belarus's *energy dependence*, one of the key factors influencing the republic's relationship with Russia and determining the format for unification. The research that has been carried out on energy issues has not focused closely enough on natural gas, instead paying more attention to the oil sector. However it is precisely natural gas upon which the Belarussian economy is crucially dependent since most of its industry is gas-fired[6].

---

grey security zone' in Jan Zielonka and Alex Pravda, eds., *Democratic Consolidation in Eastern Europe. International and Transitional Factors,* 2 vols. (Oxford: Oxford University Press, 2001), vol. 2, pp. 455-484. See also a series of articles by Ustina Markus devoted to Belarus in various issues of *Transition.*

[3] D. Mario Nuti, 'The Belarus Economy: Suspended Animation between State and Markets', in Stephen White, Elena Korosteleva and John Lowenhardt, eds., *Postcommunist Belarus* (Lanham MD: Rowman & Littlefield, 2004). See also Colin W. Lawson, *Path-dependence and economy of Belarus: the consequences of late reforms* (Paper presented at the seminar, Oxford University, Spring 2001).

[4] Steven J. Main, *Belarus and Russia: military cooperation 1991-2002* (Camberley: Conflict Studies Research Centre, Royal Military Academy Sandhurst), 2002, pp. 1-13.

[5] В. Е. Снапковский и А. В. Шарапо (ред.) *Белорусско-Российские отношения: проблемы и перспективы.* Материалы второго круглого стола белорусских и российских ученых (Минск: БГУ, 2000). See also Т. Шаклеина, *Белоруссия во внешнеолитической стратегии РФ,* Дискуссии о Союзе России и Беларуси в российском политико-академическом сообществе геополитический аспект (Москва: Московский общественный научный фонд, 2000). Also see 'О Российско-Белорусской интеграции. Тезисы Совета по внешней и оборонной политике', *Независимая газета,* 1 октября 1999.

[6] The gas share in the Belarusian energy balance is extremely high and it even exceeds the same parameter in Russia.

Belarus's 90% dependence on Russian gas imports for which it cannot pay in full and in time, together with its being an important transit route for Russian exports to Europe, made the gas issue one of the central problems of relations between the two countries, turning Belarus into a 'political flashpoint' in the process[7]. The problem is further exacerbated by the fact that the gas supply system is more capital intensive and inflexible than its oil counterpart and, once the investment in gas pipeline network is made, it is much more difficult to recover. This infrastructural inflexibility is an important factor determining Russia and Belarus, locked as they are by gas trade and transit, to look for a long-term solution as opposed to an all-out conflict.

This suggests a high degree of *interdependence* between the two countries, a feature that was somewhat overlooked in existing research on Russian-Belarussian unification, which stresses Belarus's dependence on Russia but underplays the two countries' interdependence. However, it is not only Belarus that needs gas as a consumer but it is also Gazprom that needs the Belarussian routes as an exporter. Underplaying interdependence of Russian-Belarussian relations, or, indeed, simply substituting it for dependence, is partly justified by a high asymmetry – political, economic and military - of interdependence in Russia's favour. However, such an approach at the very least fails to address the bargaining power of the weaker side and assumes that the conflict is resolved on the stronger side's terms[8]. Furthermore it leads scholars to ignore the fact that a relationship of interdependence – however asymmetrical – is costly to break for both sides. Indeed, by cutting off gas supplies to Europe Gazprom not only deprived Belarus of gas but also threatened its own long-standing reputation as a reliable gas exporter. Moreover, as C. Bruce notes, interdependence is a dynamic process[9] and therefore the cost of breaking the relationship can change over time, thus influencing both sides' bargaining power. This book will assess the cost of cutting gas off for Gaz-

---

[7]   Ewan Anderson, *An Atlas of World Political Flashpoints: A Sourcebook of Geopolitical Crisis* (London: Pinter Reference, 1993), p. 13.

[8]   James Dougherty and Robert Pfaltzgraff Jr., *Contending Theories of International Relations. A Comprehensive Survey* (London: Longman, 2001), p. 94.

[9]   Chloe Bruce, *op. cit.,* p. 3.

prom and for Belarus in political and economic terms, i.e., what it means for Gazprom as a gas supplier and Belarus as a consumer and transit country.

We will argue that both sides are interested to minimise the negative effects of the 2004 break-up and to re-establish their relations within a new framework. The old framework in which gas issues were dealt with was initially provided by the idea of Russian-Belarussian integration. Thus the book explores the history and evolution of bilateral relations in the gas sphere during the last decade in the context of 'unionist' political, economic and military tendencies, thereby analysing the gas conflict from a broader perspective than purely one of 'trade war' and arguing that the conflict was the result of a failure to establish political union between the two countries.

Importantly, Belarus and Russia attached different weight to various aspects of integration at different stages of their relationship. Whereas initially economic benefits lay at the heart of Belarus's desire to engage in the integration process, Russia attached more significance to geopolitical and military issues. However, as the economic situation in Russia worsened, Moscow reconsidered its attitude towards Belarus and started to attach more weight to geo-economics, as E. Luttwak calls it[10], rather than to geopolitics, thus shifting gas issues from the political to the commercial framework. This meant fewer subsidised prices for Belarus, which, in turn, changed its perception of the possible benefits of a union. Realising that the beneficial format might be impossible to achieve, the Belarus leadership is reluctant to advance unification on terms that would worsen the country's economic situation.

The question of the differing nature of interest in integration and its subsequent evolution on the part of Russia and Belarus is not thoroughly explored in existing scholarship but is vital to understanding why political unification failed and, consequently, unresolved gas issues resurfaced. Indeed, having initially

---

[10]   Edward Luttwak, *Endangered American dream* (New York: Simon & Schuster, 1993), pp. 307-325. See also Aron Raymond, *Peace and War* (New York: Doubleday, 1996), p. 191.

chosen a strategy of 'bandwagoning'[11], i.e. joining the stronger side to receive economic protection if at the cost of loosing some independence, Belarus became disappointed as the economic benefits began to decrease and overly serious political concessions were required to maintain them. Russia, for its part, realised that its geopolitical aims in Belarus could be achieved without establishing a formal political union. Because of these re-assessments both Russia and Belarus repeatedly violated agreements regarding political and economic integration, thus undermining its prospects since the key to cooperative behaviour lies in the extent to which each party believes the other will cooperate[12].

In Russia's perception the rewards arising from cooperation with Belarus ceased to outweigh the rewards that could be gained by unilateral action and competition, suggesting a shift from *cooperation theory to a conception of interest*[13]. In order to determine whether the same holds true for Belarus it is necessary to analyse whether Belarus is in a position to benefit (or indeed survive) from acting exclusively in a new 'commercial' framework without making political concessions. Indeed since the origins of today's gas conflict lie in Belarus's economic weakness, and its refusal to sell part of its gas transit infrastructure to Gazprom as a means of debt settlement, it is important to establish whether the Belarussian economy enables the republic to act as a genuinely independent economic actor, i.e., whether it is sustainable enough to afford paying increased prices for gas without leading to major domestic deterioration.

Existing research on the Belarussian economy is divided on the issue of whether Belarus is just a slow reformer or a non-reformer. Neither side, however, explains in detail why the largely unreformed Belarussian economy was able to sustain itself for more than a decade. The notable exception is the research conducted by M. Nuti who first addressed the issue of sustainability of

---

[11]   Paul Schroeder, 'Historical reality vs. neo-realist theory', *International Security*, 19:1 (Summer 1994), pp. 116-117.

[12]   Dougherty James and Pfaltzgraff R. Jr., *op. cit.*, p. 506.

[13]   *Ibid.*, p. 506.

the Belarussian economy, explaining it by Belarus's secured access to Russian gas at low prices[14]. This book concurs with Nuti's assessment and further argues that as soon as Russia increased gas prices the mainstay of Belarussian welfare was bound to deteriorate. To support our argument we evaluate the changes in the difference between prices that Beltransgaz paid Gazprom for gas and prices that it charged Belarussian industrial enterprises. We will show that this margin, which earlier was an important source of state budget revenues and was widely used for subsidies, decreases over times, thus undermining the basis of Belarussian 'welfare state'.

Having determined that Belarus's decision to accept higher gas prices rather than sell off its gas transit network to Gazprom was not based on substantial economic grounds, the book next address the issue of further gas prices increases. It argues that the Belarus economy's already limited ability to withstand the pressure of increased gas prices will further be weakened since Russia has increased its domestic gas prices and will continue to do so at 20% annually as part of its energy strategy and in accordance with WTO requirements. In this case, even under a joint venture between Belarus and Gazprom, the condition of gas deliveries to Belarus at Russian domestic prices, would not mean continuing gas supply at pre-2004 price level since Russia cannot ship gas to Belarus at a lower price than is paid by its domestic consumers. This means that, even if Belarus were to establish the JV, prices will be increased and Belarus will have either cut its social expenditures, an undertaking capable of triggering political instability, or start a restructuring of its highly energy-intensive gas-fired industry.

Despite its economic weakness and contrary to what the 'realist' approach would have predicted, Belarus has tried to act as an independent economic actor in its gas dispute with Gazprom and chosen to ignore the threat by refusing to establish a joint venture and arguing that the state will pay a higher price for gas rather than sacrifice ownership of the transit network. But since Belarus is, in fact, unable to pay increased prices without Russia's help, the gas dispute is not resolved on any permanent basis.

---

[14]   D. Mario Nuti, *op. cit.*, p. 121.

This book will explore whether, under these conditions, Belarus can use politi-cal and military issues as 'bargaining chips' in its dispute with Russia. We ar-gue that such attempts are unlikely to work since the level of cooperation in political and military spheres is already sufficient for Russia and it does not need any closer cooperation. Thus the economic dimension is decisive in de-termining how the gas issues will be settled.

Even if it is confined to acting exclusively in the economic dimension, Belarus still has some important 'cards' to play. Gazprom, being interested in keeping and expanding its share of the European gas market, needs to insure uninter-rupted gas transit to Europe which it cannot guarantee on an annual basis - as opposed to short-term period as in 2004 - without using gas transit routes passing via Belarus. Furthermore, Gazprom needs these routes not merely as a strategic export corridor but as a part of its export diversification strategy. While arguing that both Belarus and Gazprom are interested in returning to the negotiating table over the formation of the Beltransgaz JV, the book ex-plores how the timing of negotiations influences Belarus's bargaining power.

We will argue that the possible construction of the new offshore export pipe-line, the NEGP, which would bypass all CIS countries, will have an important effect on Belarus's bargaining power over the conditions of a JV. Apart from that Belarus's position is seriously influenced by the changes in the reliability of the Ukrainian gas transit system. Belarus and Ukraine have often been con-trasted in the literature as countries that are both vitally dependent on Russian gas but have chosen distinctly different attitudes towards acceptance of Rus-sian influence in political and economic affairs[15]. However, little attention has been paid to what these differing attitudes mean for the reliability of their gas transit routes. This book will address this issue and analyse changes in the re-liability of Belarussian and Ukrainian routes, all in the context of pro- and anti-unionist tendencies. We will explore how actual or anticipated changes in the reliability of the Ukrainian transit routes influence the importance of Belarus-

---

[15]   Ustina Markus, 'Belarus, Ukraine take opposite views', *Transition* (15 November, 1996), pp. 20-22. Also see Taras Kuzio, *op. cit.*, pp. 455-484.

sian ones, and, consequently, affect Gazprom's willingness to seek a compromise with Belarus.

This book relies on primary sources (such as Declarations, Treaties, Intergovernmental Agreements, energy news reports, interviews etc) and secondary ones. Foremost importance among the primary sources is assigned to the speeches and declarations made by the Belarussian and Russian presidents as they are by far the most influential players in the two countries' domestic political affairs. Their personal statements are most likely to demonstrate the prevalent policy conduct on important issues, including gas disputes.

The book consists of four chapters, excluding an introductory and concluding chapter, and is organised as follows.

An 'Introduction' provides a critique of existing literature, outlines the problems set and presents the main arguments of the book.

Chapter 1 'The Russian-Belarussian relationship: geopolitics and geo-economics' analyses the post-USSR evolution of the Russian-Belarussian relationship in the following dimensions – political, economical and military – and examines how the each country's geopolitical and geo-economic considerations affect the pace and format of integration as well as the means of gas conflict resolution. The chapter argues that the gas issue is shifting from the semi-political to the economic sphere, and therefore even an increasing convergence of Russian and Belarussian political systems does not make the resolution of the conflict easier.

Chapter 2 'Belarus and 'cheap' Russian gas: can Belarus afford not to reform?' analyses the current shape of the Belarussian economy, examines the degree to which its performance was relaint on previously accessible cheap Russian gas and explores its sustainability under increased gas prices. We argue that the viability of the national economy under increased gas prices can only be achieved by industrial restructuring leading to decreased gas consumption.

Chapter 3 *'Russian gas to Europe: the analysis of existing and projected export routes'* explores how the unresolved gas conflict around the Belarus gas transit infrastructure affects Gazprom's gas export routes' strategy, analyses alternative existing and projected gas export routes and argues that a sound resolution of conflictual issues with Belarus is still necessary regardless of whether the CIS by-pass route is built.

Chapter 4 *'How significant is Belarus's gas bargaining power?'* analyses how a seemingly amicable relationship between Belarus and Russia deteriorated into the February 2004 gas crisis, identifies the 'bargaining chips' that each country holds to resolve the conflict and examines how possible variants of resolution influence the reliability of Russian gas exports to Europe.

The *'Conclusion'* summarises the analytical findings of the book.

This manuscript was completed in April 2005 before the construiction works on the onshore section of the NEGP had begun and before the January 2006 Russia-Ukraine gas crisis had happened. These events have further highlighted the importance of problems of transit of Russian gas examined in this book.

1   The Russian-Belarussian Relationship: Geopolitics and Geo-economics

1.1   The Break-up of the USSR and the Creation of the CIS: Perceptions and Interests

This book begins with an exploration of the roots and currents of Russian-Belarussian relations and the evolution of the idea of the union state since the dissolution of the USSR, including such stages as the creation of the CIS, the Customs Union, the CSR and the Union of Russia and Belarus. It explores the nature of the interests that each country holds in one type of union or another in order to evaluate perspectives of each format and their implications for the reliability of the Belarussian gas transit routes.

From the moment of the declaration of independence in 1991 there was never a strong assumption in Europe that Belarus would become a member of the 'fourth wave' of democratisation, following the east European countries. Further events proved that such scepticism was well grounded[16]. This book argues that the major initial factor that impeded the pursuit of democratisation in Belarus was the failure of the national idea to unite its population[17]. One of the important factors that limited popular support for the Belarussian National Front, BNF, was its ultra-nationalist nature, a feature that did not have much appeal for a largely russified Belarussian public. By contrast, in Ukraine, where the national idea proved to be a major force pushing towards democratisation, the national movement was very diverse and included both ultra- and moderately nationalistically-minded elements, thus providing for a larger base of popular support.

---

[16]   Alexander Lukashuk, 'A year on a treadmill', *RFE/RL Research Report,* 2:1 (January 1993), p. 68.

[17]   Taras Kuzio, *op. cit.,* p. 456. For disculssion of the role of the national idea in democratisation see Dankwart Rustow 'Transitions to democracy: toward dynamic model', *Comparative Politics,* 2:3 (April 1970), p. 350.

Importantly, unlike in other Soviet republics where it initially strengthened national movements, a dialogue between the genuinely national-democratic
forces and communist nomenklatura, based on the primacy of the national
idea, was largely absent in Belarus[18]. This can be attributed to the fact that the
nomenklatura did not regard the national idea as a safe bet and therefore
made no attempts to coalesce with the nationalist forces. This can also be explained by the absence of 'soft-liners' in the Belarussian communist party
(BCP) leadership who would have been willing to have such a dialogue. Indeed, the BCP elite was one of the most reactionary among the USSR republics not having produced single 'dissenting' communist, unlike, for example,
the all-Union CP. Vivid evidence of the strength of Belarussian 'hard-liners', as
K. Mihalisko rightly observes, is that the republic was the only one in the
USSR that, in the first multi-candidate elections to the national legislature,
chose to reserve seats for conservative forces instead of making all the places
available for an open contest, gaining a 'Vendée of perestroika' label in the
process[19].

Only in 1991, when the coup d'état discredited itself, did the BCP begin to 're-
vert itself into national communism', to 'legitimise' its authority in the 'national
context'[20]. It broke with the CPSU and provided for passing the Declaration of
Independence in the national legislature. This move culminated in the joint
signing of the Belovezha Accord with the Russian and Ukrainian presidents -
its chef initiators – and creating the CIS. However, playing the national card
belatedly, the Belarussian conservative leadership did not gain popular support and only further alienated its remaining supporters. According to the polls,
the majority of the Belarussian population disapproved of the Belovezha
agreement lending 83% support to the 'revived' USSR (curiously, the same
proportion supported economic integration with Russia four years later). This
figure, as T. Kuzio notes, was the highest outside Central Asia, and higher still
than the all-Union average[21]. Thus, given a popular referendum, Belarus

---

[18]   Kathleen Mihalisko, *op. cit.*, p. 240.
[19]   Kathleen Mihalisko, *op. cit.*, p. 239.
[20]   *Ibid.*, p. 241.
[21]   Taras Kuzio, *op. cit.*, p. 477.

would not have supported the dissolution of the USSR. This contrasts with Ukraine, where a referendum actually took place and got 90% of the vote at an 84% turn-out.

At the same time, had the Belarussian leadership not signed the Belovezha Accord, it would have alienated itself from Yeltsin's Russia. This book contends that the fear of losing Russia's support was one of the important factors influencing Belarus's actions since an advocated closeness to Russia, as opposed to the national idea, was seen as a major platform capable of uniting the Belarussian public and delivering political dividends.

Indeed, Moscow's support was perceived as almost guaranteeing a victory in the first presidential elections in 1994[22]. This, however, soon proved wrong when the pro-Moscow candidate, Kebich, lost out to the virtually unknown independent candidate, Lukashenko. Thus in view of the weak national consciousness and growing disappointment with the incumbent leadership, signing the Belovezha Accord substantially alike decreased the chances of both the incumbent leadership and nationalist forces, thus creating a political clearing for a candidate whose most prominent campaign message was his disapproval of the Belovezha agreement. Not surprisingly, when Lukashenko came to power his unifying idea was not so different from Kebich's – building stronger links with Russia, both economically and politically while at the same time undertaking only very limited reforms domestically.

A desire to have Russia's support is an important characteristic feature of the majority of Belarussian political forces[23] (apart from the BNF whose influence was rather weak), which perceive closer economic integration with Russia as being economically beneficial, if not vital for Belarus. This book argues that such a perception initially gave a strong stimulus and went on to shape the subsequent unification processes between Belarus and Russia. Further impetus towards integration was given by a strong fusion of Soviet and national

---

[22]   *Народная Газета,* 17 Июня 1994.

[23]   Taras Kuzio, *op. cit.,* p. 476.

elements in the identity of the Belarussian population, which perceives an alliance with Russia as a restoration of Soviet identity.

In Russia itself opinion polls were in favour of the dissolution of the USSR only for a very short time, at the peak of Yeltsin's popularity after the failed coup, only to shift towards disapproval a few months later. Since then opinion polls have consistently suggested a high degree of approval for the integration processes with Belarus[24].

A distinct difference between Belarus's and Ukraine's attitude towards the CIS, was that Belarus saw it as a basis for closer integration whereas in Ukraine it was perceived as 'a civilised divorce'[25] and a way to gain more independence later on. This explains Russia's initially different attitude towards both countries in the sphere of energy[26]. Because of Ukraine's firm stance regarding its independence, the Ukrainian gas transit routes were seen as *a priori* unreliable (which was to be proved later), whereas the Belarussian ones were seen as absolutely secure due to Belarus's political 'closeness' to Russia. Therefore, as a result, the Russian gas exports monopoly Gazprom undertook further significant investments in transit routes passing through Belarussian territory, i.e., Yamal-Europe pipelines system. However, as the 2004 gas conflict demonstrated, the Belarussian routes proved to be no more reliable than the Ukrainian ones, thus making Gazprom seriously consider the construction of an alternative route by-passing all CIS countries. Did such deterioration happen unexpectedly, or were there some underlying problems in the seemingly smooth Russian-Belarussian relationship? In order to answer that, this book will next address the details of the evolution of the integration process between the two countries over the last decade.

---

[24]   62% in support in 1997, 72% - in 2000. Richard Sakwa, *Putin: Russia's Choice* (London: Routledge, 2002), p. 232. Also 80% in 2001 according to *RFE/RL* (22 August, 2001).

[25]   Alex Pravda, 'Russia and the 'near abroad' in *Developments in Russian Politics 5*, in Stephen White, Alex Pravda and Zvi Gitelman, eds. (London: Palgrave, 2001), p. 16.

[26]   Ustina Markus, 'Belarus, Ukraine take opposite views', *Transition* (November 1996), pp. 20-22.

1.2   The Russian-Belarussian Integration:
Three-dimensional Evolution

The cooperation between Russia and Belarus in the three different dimensions – political, economic and military - has been uneven with some dimensions progressing fast and others facing serious impediments. This book explores the nature of the unresolved problems of integration in each dimension and analyses their influence on Russian-Belarussian relations in the gas sphere, particularly on the nature and timing of the gas conflict resolution.

Political Dimension

As noted earlier, due to a heavy economic dependence on Russia and an aversion to market reforms, there was a strong initial drive on the part of the Belarussian leadership, following the break-up of the USSR, to preserve close ties with Russia, when the CIS was seen as a step towards further integration rather than a move towards complete independence.

Russia, on the other hand, having difficulties in perceiving itself a separate entity from the former republics-turned-independent states, largely considered the CIS as the initial ground to keep its influence and possibly re-gain its dominance over the former republics[27]. However, such perception of the CIS was not prevalent from the moment of its creation. Indeed, for Yeltsin, the CIS was rather a convenient way to break with the USSR and the central authority, and, on more personal terms, with Gorbachev. Only later, when, as M. Light notes, Russia realized that the post-USSR's 'external environment proved less benevolent' than expected[28], did Russia begin to perceive its national interest in terms of acquiring more influence within the CIS.

---

[27]   Margot Light, Post-Soviet Russian Foreign Policy: The First Decade in Archie Brown, ed., *Contemporary Russian Politics: A Reader* (Oxford: Oxford University Press, 2001), p. 422.

[28]   Margot Light, *op. cit.*, p. 420.

After some time, however, it became clear that the Russian state was too weak economically to undertake the full-fledged integration of the CIS, and that the CIS could not be used as 'an instrument for the Russian aggrandisement policy'[29]. Instead, as R.Allison notes, Russia chose to develop 'genuine' bi-lateral partnerships with those states which it perceived as most important for securing its national interest[30]. Due to a number of geo-strategic considerations - having Russia-owned export oil and gas pipelines running through its territory is one of them - Belarus *is* an important partner for Russia. Belarus's provision of sites for the Russian military, participation in common air defence and serving as a transport corridor to Russian Kaliningrad further underlines the importance of mutually good relations between Russia and Belarus[31].

Therefore, despite being disappointed at the failure of the pro-Moscow candidate and despite initially refraining from entertaining close connections with the newly elected Belarussian president, Russia soon resumed the talks regarding further integration. Importantly, Russia's intentions to establish closer political, economic and military links with Belarus met a highly positive response from the new Belarussian leadership[32]. Such an attitude was favourably contrasted with Ukraine's enduring reluctance to join any Russia-led CIS security structures, a desire to keep its non-alliance status and an intention to develop a 'Euro-Atlantic' foreign policy orientation, thus making Belarus a yet more important partner in the strategic sense[33].

Thus, starting from September 1995, the process of integration speeded up, especially in the military and security spheres – with Yeltsin's decree 'On the

---

[29] Richard Sakwa, *op. cit.*, 2004, p. 211.
[30] Roy Allison, Russia and the New States of Eurasia in Archie Brown, ed. *Contemporary Russian Politics: A Reader*, p. 444. See also T. Шаклеина, *op. cit.*, pp. 13 and 21-34 for an idea of 'selective integration'. See 'Концепция внешней политики Российской интеграции' (утв. Президентом РФ 28.06.2000) which emphasises the shift from multilateralism to bilateralism in relations with the CIS in И.Иванов *Новая Российская дипломатия* (Москва: ОЛМА-ПРЕСС, 2002), стр. 210-232.
[31] Steven Main, *op. cit.*, p. 8.
[32] T. Шаклеина, *op. cit.* p. 56.
[33] *Ibid.*, p. 43.

establishment of the strategic course of the Russian Federation with CIS member-states'[34], which marked a new stage in Russian foreign policy, often referred to in the literature as that of 'a great power rhetoric'. This book examines how much of the integration process that followed was due to rhetoric, and how much to real action.

The first talks regarding political integration with Belarus commenced in 1995 during Yeltsin's visit to Minsk, when the Treaty of Friendship, Good Neighbourliness and Cooperation was signed. In April 1996 Russia and Belarus founded the Community of Sovereign Republics (CSR)[35]. The integration process intensified in early 1997 when, in a letter to Lukashenko, Yeltsin declared the necessity to achieve 'the fuller implementation of the integration programme'. Moreover, Yeltsin accepted Lukashenko's original proposal to hold a referendum in both countries on the issue of unification and formation of a common government. In the event a vote did not take place but a treaty about creation of the Union of Russia and Belarus was nonetheless signed in April 1997 and ratified by strong majorities in both legislatures[36].

In essence, the new structure was close to confederation since it envisaged 'adopting a common foreign policy, the creation of a ruling Supreme Council, including both heads of state and government, the establishment of the executive committee and a joint parliamentary assembly'. Seemingly, more mechanisms were being put in place in order to allow for a greater coordination between the two countries. However, both the Union's legislature and executive were largely decorative organs and dependent on the executive/legislature of their respective member-states. Reportedly, the Belarussian side pushed for a more powerful Union's executive and legislature. This move, however, was al-

---

34 *Российская Газета*, 23 Сентября 1995.

35 В. П. Снапковский, 'Сценарии белорусско-российской интеграции' in *Белорусско-Российские отношения: проблемы и перспективы (*Минск: БГУ, 2000), стр. 44.

36 In both chambers of the Russian legislature - 363/2 and 144/0. In Taras Kuzio, *op. cit.*, p. 467.

legedly opposed by Chubais, the neo-liberals and the oligarchs[37]. In the end the project was re-written in order to prevent the Union's organs from having powers more extensive than those of a Russian President. Changes in the Russian domestic political landscape had a profound effect on the intensity of integration. Indeed, when 'the young reformers' fell out of power, the integration process intensified again, marked by the signing in December 1998, of The Declaration about further integration of Belarus and Russia. A year later, in December 1999, Yeltsin and Lukashenko finally signed the Union treaty in Moscow[38].

However, in 2000, with Putin's accession to the presidency, integration processes slowed down when he refused to accept an integration formula endorsed earlier by Yeltsin. Putin's position was that the unification must proceed 'unconditionally' on the basis of a single state with a single parliament and a single government[39]. Thus he spoke out against creating a Supreme Council envisaged in 1999 treaty calling it a "supranational organ with undefined functions"[40], and Belarussian proposal of a constitutional act as being altogether a 'legalistic nonsense'. Putin proposed to hold a referendum on unification in May 2003, elect a joint parliament in December 2003, introduce Russian currency as Union's currency from January 2004, and elect a new Union President in 2004[41]. In fact, such an offer envisaged the incorporation of Belarus into the existing Russian Federation rather than the creation of a new federation. Lukashenko rejected the offer calling it 'even insulting'[42] and aiming at stripping Belarus of its sovereignty. Still the uncertainties concerning the format for union persisted, with an agreement reached in 2003 on the basic provisions of the Constitutional Act of the Union, envisaging a separate union leg-

---

[37]   'Россия-Беларусь от союза к союзному государству' in *Белорусско-Российские отношения: проблемы и перспективы* (Минск: БГУ, 2000), стр. 35-43. Also see Bobo Lo, Russian foreign policy in the post-Soviet era: reality, illusion and mythmaking (London: Basingstoke, Palgrave Macmillan, 2002).
[38]   *Независимая Газета*, 9 Декабря 1999.
[39]   *RFE/RL*, 24 June 2002.
[40]   *RFE/RL*, 13 June 2002.
[41]   *RFE/RL*, 14 August 2002.
[42]   Chloe Bruce, *op. cit.*, p. 13.

islature, and a government that would interact with two sovereign nations[43]. However, an exact mechanism was not specified, leaving it unclear whether two countries were to keep their respective legislatures and executives, or introduce a single authority instead.

The dispute over the format of political Union can be seen as a choice between three scenarios of integration – confederation, the new Russian-Belarussian Federation, and Belarus's incorporation into an already existing Russian Federation[44].

This book examines each country's attitudes towards these different scenarios. It appears that the first variant is the one most supported by the Belarussian government, whereas Russia clearly rejects Yeltsin's confederation formula and pushes, instead, towards the third variant. There are several explanations for Russia's position. Firstly, in the event of including Belarus in the Russian Federation, Russia would not need to re-apply for permanent membership of the UN Security Council in order to transfer it to the Union state. Secondly, there is a smaller danger of triggering domestic instability in Russia itself by including Belarus as yet another region rather than creating a new federation. In the latter case there is a danger that some Russian national republics, such as, for example, Tatarstan, might demand to upgrade their status within the new federation to achieve more independence, thus contributing to Russia's 'asymmetrical federalism'.

For the Belarussian leadership, however, the third scenario seems to be totally unacceptable since it implies a loss of independence resulting in a substantial reduction of presidential bureaucracy – both the presidential 'vertical' administration and the security council - thus making it fiercely resist the scenario. Since these two institutions constitute the president's main base of support, he is careful not to alienate them[45]. Another reason for this resistance is

---

43   *RFE/RL*, 11 March 2003.
44   Ю. П. Бровка, 'Белорусско-Российская интеграция с точки зрения международного права', in *Белорусско-Российские отношения: проблемы и перспективы* (Минск: БГУ, 2000), стр. 29-34.
45   Kathleen Mihalisko, *op. cit.*, p. 257.

of an economic nature since the 'incorporation' format would significantly reduce the ability of the government to adopt independent economic policies
thus threatening the current political regime.

Since Belarus prefers confederation whereas Russia opts for including Belarus into the existing federation, there are not many common points left in the
two sides' positions regarding the format of political unification. Therefore, this
book argues that the prospects of the Russian-Belarussian political union any
time soon are rather slim.

The book contends that, ironically, an absence of agreement regarding the
type of political union persists despite the growing convergence of the Russian
and Belarussian political systems. Indeed, since Putin's accession to the
presidency the Russian political system has been developing along more authoritarian lines than it had been the case under Yeltsin, and in some respects
it has been converging towards the Belarussian political model. For example,
the recent presidential 'vertical' building federal reforms in Russia such as the
institution of governors-general, replacement of governors' direct elections by
appointments, right of the president to dismiss regional legislatures, are reminiscent of the steps taken earlier by the Belarussian president. However, such
convergence of the two countries' political systems does not imply that Russia
is willing to support a largely unreformed Belarussian economy.

Therefore, a union where Belarus would be 'together with Russia' as its equal
political partner is not really an option since Russia refuses to accept it[46].
Growing 'economisation' of Russia's foreign policy suggests that Belarus
faces a choice between being 'together with Russia but politically subordinate
to it', or a separate, politically independent entity, with all cooperation being
done on purely economic grounds. This book contends that since economic
factors gained prevalence in Russian-Belarussian relations, replacing political
ones, these greatly influence the type of union, or indeed the absence of *any*
political union, that both parties would agree upon.

---

[46]   *RFE/RL*, 11 June 2002.

While T. Kuzio is right to argue that geopolitics superseded Russia's political considerations about democracy in Belarus[47], we further argue that geopolitics is unlikely to supersede Russia's economic considerations. Indeed, Russia, even if willing, is not economically capable of supporting Belarus further irrespective of whether or not there is a political union. Thus 'muddling through', i.e., in effect, doing nothing to make political integration real, could continue for years to come unless the economics of bilateral relations is resolved. Therefore this book contends that if geo-economics would converge with geopolitics the union would be strongly endorsed and supported by Russia. But as the situation stands now – when geopolitical considerations cost Russia dear in economic terms - geo-economics prevails.

However, the book emphasizes that Moscow cannot 'afford' to forego its geostrategic interest in Belarus and 'wash its hands' if Belarus does not implement economic reforms. We argue that since Russia realizes the geo-political importance of Belarus, and even more so in the wake of the Ukrainian political reforms, it is willing to help Belarus to become a more valuable economic partner, not by subsidizing its economy as before, but rather by imposing such conditions on Belarus that would force its leadership to embark on reforms out of economic necessity. Indeed, an economically stronger Belarus is also a better ally for Russia in a geopolitical sense. Moscow thus can play an important role in making its geopolitical and geo-economic attitudes toward Minsk converge.

The book further examines what the various union formats imply in terms of the gas trade[48]. The federation scenario means that the ownership and management of the gas transit infrastructure would be transferred to the federal centre. Further developments would differ depending on the type of federation. In the event of Belarus's incorporation into the Russian Federation, the management of the Belarussian transit network would automatically be transferred to Russian federal structures thus becoming Russian state-owned with

---

47   Taras Kuzio, *op. cit.*, p. 468.
48   'Газовые приоритеты Российской дипломатии', *Дипломатический Вестник*, Февраль 2000.

a possibility of its eventual resale to the Russian companies, most naturally to Gazprom. In the event of the creation of the new, Russian-Belarussian, federation, the ownership and management would be transferred to the new federal centre. Under these two options Belarus's ability to influence conditions of gas system management and, furthermore, the amount of transit fees, are either non-existent or very limited. Confederation status implies joint management of the gas transit system, and therefore allows more room for negotiations on the Belarus side than the federation scenarios.

Since the integration process has been going on for years and there is still no common state in any form, we should not forget a fourth option, that of Belarus remaining an independent state, not participating in any of the aforementioned political unions. In this case, Belarus would keep its gas transit system network. However, how long it would retain it depends on whether Belarus is able to sustain itself economically under the pressure of increased gas prices without being forced to sell off its network to Russia. This book contends that Belarus could lose full control over its gas transit system not only if it became a member of a union state, but also if it remained an independent but economically unviable state. The question of economic sustainability is discussed in Chapter 2.

Economic Dimension

The establishment of the Customs Union (CU) in 1995 was the first step towards closer economic integration – a process approved by an absolute majority of the Belarussian population, 83%, in the 1995 referendum[49]. The CU offers important benefits to both countries such as free access to each other's markets, cuts in expenditure for border maintenance, and simplification of contacts in the military-industrial complex[50].

---

[49]   В. П. Снапковский, *op. cit.*, p. 54.

[50]   А. В. Шурубович, 'Проблемы и перспективы российско-белорусских экономических отношений' in *Белорусско-Российские отношения: проблемы и перспективы* (Минск: БГУ, 2000), стр. 69-80.

However, the CU Agreement is still not implemented in full and is regularly violated by both countries. For example, Russia and Belarus equalized their import taxes as required by the CU Agreement, but soon afterwards both started to change them unilaterally[51]. Belarus further violated the agreement by imposing limits on its cheaper agricultural production exports to Russia.

The most difficult component of economic integration proved to be that of common currency, with the two parties repeatedly failing to agree on currency and on a central money-issuing facility[52]. Belarus wishes to keep its own central bank, whereas Russia insists on a single central bank in Moscow. In mid-2003 both sides agreed to introduce the Russian rouble as a single currency from January 2005, and a common currency from January 2008. This agreement envisaged both countries retaining their central banks but, in fact, transferred crucial functions such as money-printing, exchange-rate setting and the development and implementation of monetary policies to the Russian central bank. Belarus was permitted to carry out transactions within limits proportional to its GDP (4.2-4.3 % of the combined GDPs of Belarus and Russia). At the last minute, however, Minsk refused to agree on the single emission centre in Moscow, arguing that it would signal the end of Belarussian independence[53]. This book argues that the prospects of establishing a common currency are further weakened by Russia's making its introduction conditional on the resolution of the disputes surrounding the Beltransgaz JV.

Furthermore, gas issues settlement is also one of the preconditions for the Common Economic Space, a multi-lateral economic union between Russia, Belarus, Ukraine and Kazakhstan, which Russia launched in 2003[54] and in which it showed an increasing interest as bi-lateral economic union with Belarus came to a standstill. While 'economising' its foreign policy towards Belarus, Russia's attitude towards the CES as a whole has to do as much with

---

[51]   Ibid., стр. 73.
[52]   RFE/RL, February 23 2000. See also John Odling-Smee, Monetary union between Belarus and Russia: an IMF perspective (Washington, DC: IMF, 2003).
[53]   Public support in Belarus for common currency declined from more than 50% in 2003 to 34.4% in 2004.
[54]   RFE/RL, February 23 2003.

politics as it has with economics. Indeed, the CES can be seen as yet another attempt to engage Ukraine, which earlier refused to participate in another multilateral economic union, the Eurasian Economic Commonwealth. Getting Ukraine into the CES would allow Russia to have more room for manoeuvre when it comes to negotiating gas transit via Ukraine. The book argues, however, that it is too early to assess the prospects of the participation of Ukraine in the CES in view of the recent political changes in that country.

The cooling down of relations between Russia and Ukraine might give a fresh impetus to Russian-Belarussian economic union. However, the major reason for the earlier poor progress in two countries' economic unification, i.e., the fundamental difference in their economies – Russia moving towards a market economy and, in essence, Belarus remaining a command economy - is still in place. Thus differing levels of economic development, industrial structure, dependence on foreign trade and energy resource endowments weaken unification prospects. The Belarussian economy's high dependence on Russian gas, coupled with its inability to pay for it in full, and at the same time its reluctance to establish the JV, which Russia insists upon as a pre-condition of closer economic union, further impede economic unification.

Military Dimension

Since declaring independence, Belarus's stated aim is to become a neutral state. However, from the start there was an internal conflict between the national executive and the legislature – with the latter pushing towards the CIS collective security treaty, and the former – towards neutrality and non-aligned status. The executive prevailed and in less than a month from the establishment of diplomatic relations between the two states Belarus and Russia had signed 5 military agreements dealing with the countries' strategic and conventional forces. The initial integration drive was greatly hastened by the common perception of NATO's eastward expansion as a threat, and also by the fact that Ukraine showed an interest in cooperation with NATO and did not participate in the creation of any effective defence network within the CIS.

This book argues that from the outset the military component of Russian-Belarussian integration was very important for the incumbent Belarussian leadership, which, unlike other CIS countries, had not diversified its traditional military links and sought to secure Russian protection in the wake of its close neighbours' – Poland and the Baltics – desire to join NATO. Furthermore, Belarus insisted on the signing of a formal military alliance with Russia as an insurance of Belarussian security[55] despite the fact that the Charter of the Union of Russia and Belarus already envisaged an obligation to take 'joint measures to avert a threat to the sovereignty of each member'.

Having stressed the importance of military cooperation with Russia for Belarus, let us now examine what Belarus, as a member of a military alliance, signifies for Russia. Undoubtedly, Belarus is important in geo-strategic and military terms, first of all because of its geographical position in Europe, as a result of which Belarus plays an important role in the defence of the Russia's western regions, especially Kaliningrad oblast, which hosts the Baltic Fleet, and also in maintaining two Russian military strategic sites at Gancevichi and Vileika, ensuring radar cover and enabling communication with Russia's strategic submarine fleet. In 1995 Belarus granted Russia the right to use both sites for a minimum period of 25 years as well as to construct an early-warning radar site as part of Russia's plan to develop a mobile early warning attack system[56]. Thus having Belarus as a military ally enabled Russia to fill an important gap in its strategic 'security fence' which appeared when Russia lost its military bases in the Baltics.

By the end of 1998 Russia and Belarus had created a regional military security sub-system within the overall framework of the CIS collective security system and a unified regional PVO system. Finally the two countries adopted a common military doctrine – the joint Union Military Doctrine, - which, in fact, repeats the Russian Military Doctrine with minor amendments regarding nu-

---

[55] *RFE/RL*, 22 April 2003. The Belarussian President during his annual address to the National Assembly stated that only a "political, military and strategic union" with Russia could ensure Belarus's security.

[56] *RFE/RL*, 3 October 2001.

clear strategy[57]. It is thus another argument of this book that military cooperation between Russia and Belarus advanced faster that the cooperation in the
political and economic dimensions analysed above[58].

However, as S. Main correctly states, even if military integration is very close,
the steps taken so far fall short of full military union[59]. Indeed, even given that
the Russian side does not underplay the importance of Belarus in a strategic
sense, it nevertheless makes a distinction between the military alliance with
Belarus and having a common army. The Union military doctrine foresees the
creation of a Russian-Belarussian unified military grouping under a single
command, but this is seen as a possibility - 'if and when a suitable threat
arises' - rather than an imminent reality. Nonetheless, the Belarussian side
tried to attract much publicity to this grouping, making it look like a real army,
which suggests that Belarus is as much or more interested in military cooperation than Russia.

Russia stresses that the current level of military co-operation, with the two ar-
mies remaining separate entities, fully meets Russian security demands[60].
This book agrees with K. Mihalisko's argument that Russia's unwillingness to
entertain the possibility of closer co-operation with Belarus is also connected
with the Belarussian President, who feeling secure in his military alliance with
Russia, on several occasions made very harsh anti-West and anti-NATO
statements proving to be rather too erratic a partner for Russia and leaving the
Russian leadership increasingly embarrassed about its closest ally[61].

Even in the wake of September 11th, as Russia underwent a re-evaluation of
threats and concluded that an 'international terrorism' was a more important

---

[57]   *RFE/RL*, 29 August 2001.
[58]   Ruth Deyermond, 'The State of the Union: Military Success, Economic and Politi-
        cal Failure in the Russia-Belarus Union', *Europe-Asia Studies,* 56:8 (December
        2004), pp. 1191-1205.
[59]   Steven Main, *op. cit.* p. 11.
[60]   *RFE/RL*, April 21 2000.
[61]   Kathleen Mihalisko, *op. cit.*, p. 275.

security problem than NATO's eastward expansion[62] thus weakening the drive towards military unification with Belarus, Russian-Belarussian co-operation remained important on a regional basis. Also bearing Russia's weakened position in Ukraine in mind, it is legitimate to assume that Russia would not risk alienating its only remaining ally on the western frontier. At the same time, that would not make closer military integration, such as having a joint army, necessary.

Therefore the book contends that since the current level of military integration is adequate for the security needs of the both countries, and is not less important for Belarus than for Russia, the attempts of the Belarussian government to use military issues as the 'bargaining chips' in the gas conflict are bound to fail.

---

[62]   Richard Sakwa, *op. cit.*, p. 220.

2   Belarus and 'Cheap' Russian Gas: Can Belarus Afford Not to Reform?

The Belarussian leadership has long perceived Russia as a source and guarantor of the non-deterioration of the republic's living standards, which is correctly believed to be a major condition of the incumbent regime's stability. Thus the primacy of economic benefits for the Belarussian public[63] has an important impact on the political economy of Russian-Belarussian unification, explaining why the Belarussian leadership has always stressed that integration would have solved the majority of Belarus's economic problems[64].

The 1998 Russian financial crisis, causing a sharp decrease in Belarussian exports to Russia, followed by Belarus failing to find alternative export markets both because of low competitiveness and political tensions with the West, strengthened Belarus's perception of Russia as being its irreplaceable economic partner. However, the crisis simultaneously changed Moscow's attitude towards Belarus. If, as we argued earlier, Russia initially attached a higher importance to geopolitical considerations as the main driver for closer integration with Belarus, after the crisis it came to realise that its economy could no longer sustain its aspirations to superpower status. Consequently, Russia began to attach greater weight to economic aspects of integration as opposed to the geopolitical ones. With Putin's accession to power this trend solidified. He stated that unification would not proceed 'at the expense of Russian economic interests'[65], and that political and economic issues regarding Belarus will be dealt with separately.

---

63   Interax, 24.05.04. According to Belarussian public opinion centre NOVAK economic interests prevail among the Belarussians being the main reason for integration.
64   Kathleen Mihalisko, *op. cit.*, p. 248
65   *RFE/RL,* June 13 2002.

Evidence of important changes in Moscow's attitude towards Belarus came shortly later when an Intergovernmental Agreement conditioned continuing gas supply to Belarus at domestic Russian prices on Gazprom's acquisition of a stake in the Belarussian gas transit network, Beltransgaz. Furthermore, when the joint venture that had been agreed upon was not established by 2004, Gazprom, acting with the consent of the Russian government[66], demanded the Belarussian government to pay higher gas prices and, importantly, started to insist on actual payments. This is a clear contrast with the situation in 1997, when Gazprom continued to ship gas to Belarus at low prices while disconnecting its domestic non-paying consumers[67].

The consistent 'economisation' of Russian foreign policy towards Belarus in turn changed the perception, on the part of Belarussian officialdom, of the possible economic benefits that union with Russia might bring. This book argues that Russia's role in securing the Belarussian regime's stability underwent a serious re-assessment when a union was *no longer seen as necessarily being economically beneficial* for Belarus. In the eyes of the Belarussian leadership economic pressure, imposed by Russia through an increase in gas prices, undermined the very logic of the desirability of union. Indeed, the regime regards the structural reforms necessary to enable the national economy to pay increased prices for gas as capable of causing sharp welfare deterioration and thus the regime's instability. These fears are grounded in the fact that 30% of Belarussian enterprises were loss-making in 2004[68] (a steep increase from 8.8% in 1999) thus making the massive shutdowns and layoffs resulting from reforms inevitable.

The perception that the regime's stability depended on the non-deterioration of the economic situation explains the high sensitivity of the Belarussian government about entering any unification agreements with Russia capable of

---

[66]   'Russia now threatens a Belarus by-pass'. *Gas Matters* (October 2003).
[67]   Jonathan Stern, 'Soviet and Russian gas: the origins and evolution of Gazprom's export strategy', in Robert Mabro and Ian Wybrew-Bond, eds., *Gas to Europe. The Strategies of the Four Major Suppliers* (Oxford: Oxford University Press, 1999), p. 138.
[68]   *Transition Report 2004* (London: EBRD, 2004).

making Belarus worse off economically, despite its long-standing rhetoric of Slavic unification. Thus integration progresses at a much slower pace, and, as opinion polls suggest, the slowdown appears to be positively perceived by the Belarussian public[69].

In turn, Belarus is now also placing stronger emphasis on the economic dimension, which explains, for example, the occurrence of such incidents as its attempt to charge Russia for PVO service and for the two Russian military sites in Belarus in violation of previously achieved agreements. However, as we have shown before, such attempts are bound to fail since Belarus is no less interested than Russia in military cooperation. Thus there is a well grounded and growing belief in Russia that, in fact, it can get all it needs from Belarus without actually establishing a union of *any* type and simply by applying its vast economic and political leverage.

Thus the book contends that the idea of unification underwent serious downgrading in both Russia and Belarus, though for different reasons. For both countries economic benefits become the main criteria for choosing the format and timing of unification, or, as the case may be, non-unification. The partnership, justly called by R. Allison 'an alliance of convenience'[70], ceased to be such as the economic situation worsened, departing from one based on geopolitical considerations to one based on economic pragmatism[71].

Belarus has already tried to act as an independent economic actor in its dispute with Gazprom, refusing to establish a joint venture and arguing that the state would be better off paying a higher price for gas rather than losing its ownership of the gas transit network. Furthermore, Belarus threatened to increase transit fees for Russian gas flowing through the Belarussian network, if Gazprom increased its prices. However, the question that requires careful

---

[69]   Accordingly to BelaPAN, public support for economic union declined from 83.3% in 1995 to 54% in 1999.

[70]   Roy Allison, 'Russia and the new states of Eurasia' in Archie Brown, ed., *Contemporary Russian Politics: A Reader,* p. 451.

[71]   Margarita Balmaceda, 'Myth and reality in the Belarussian-Russian relationship', *Problems of Post-Communism,* 46:3, 1999, pp. 3-13.

analysis is whether Belarus actually *is* an independent economic actor, i.e., whether its economy is sustainable enough to afford paying increased prices for gas without triggering a major deterioration of the domestic economic situation. Indeed, Belarus's financial ability to withstand price increases is precisely what determines the limits of its bargaining power in its dispute with Gazprom.

2.1   An Analysis of the Belarussian Economic Situation

This book begins its examination of the sustainability of the Belarussian economy by describing it using EBRD transition indicators. Despite being somewhat subjective[72], they provide a clear picture of the progress of reforms in the
country. Presented in Table 2.1 (see Appendix) comparative trends of the
Belarussian and Russian transitions suggest that the degree of 'marketization'
of two countries differed substantially, with Russia moving towards market
economy while Belarus 'hardly started its journey'[73]. As a result, their real
economic variables such as real wages, labour productivity, unemployment
and FDI trends differ substantially. The level of reforms in Belarus has been
between very low or non-existent, as opposed to Russia, which, despite all its
failings, scores consistently higher in all assessed categories.

The only areas in which Belarus scored modestly optimistically are price and
trade/foreign exchange liberalisation. However, these policies were soon reversed by the re-introduction of restrictions on foreign exchange convertibility
and price controls, and regulations over 'socially significant goods' - whose
share in total transactions in 2003 constituted 21% - thus bringing shortages
and decreasing enterprises' profitability.

The 1993 large-scale privatisation program was suspended in 1996 and has
not advanced further since then. Its main emphasis remains limited to corporatisation, i.e., turning enterprises into joint stock companies. Thus the share
of output generated in the Belarussian private sector constitutes only 25% of
GDP, the lowest figure among all transition states[74]. Even that evaluation is
believed to be over-optimistic since it is based on data for corporatisation that
does not necessarily imply change of ownership. For example, the national
gas transit and distribution company, Beltransgaz, was turned into a joint

---

[72]   'Overoptimistic' according to M. Nuti, *op. cit.*, p. 111.
[73]   D. Mario Nuti, *op. cit.*, p. 121.
[74]   *Ibid.*, p. 113.

stock company in May 2002 but still remains state-owned[75]. *De novo* private firms in Belarus constitute a tiny proportion to GDP due to serious obstacles to entry.

Due to the virtual absence of large-scale privatisation and imposition of hard budget constraints, the structure of Belarussian industry is still not too different from that of the Soviet times, i.e., dominated by large enterprises. This further adds to the government's reluctance to undertake industrial restructuring since it would lead to massive layoffs capable of triggering regime instability. The domination of the banking system by state-owned banks (80% in 2004[76]) enables the government ordering them to lend to the loss-making enterprises to avoid mass layoffs. The inefficient industrial structure is further maintained by substantial state subsidies, which in 2004 stand at 18.9% of the GDP.

The Belarussian economy is heavily dependent on foreign trade with some 50% of its GDP coming from exports. Belarus exports 'agressively', as I. Mikhailova-Stanyuta argues, because of the economy's desperate need for hard currency[77]. Thus many of Belarussian exports are ineffective. Exports to Russia are the least effective of all since they are characterised by a high share of barter and credit, and many of them are connected with payments for Russian gas imports. However, since Belarus did not diversify its foreign trade, or rather failed while trying to do so for political and economic reasons, half of its exports still go to Russia. At the same time economic interdependence between the two countries highly asymmetrical given that trade with Belarus constitutes only 7% of Russian foreign trade. Being the largest market for Belarussian goods provides Russia with a huge leverage over the Belarussian economy (and politics) since a loss of the Russian market niche would tremendously weaken the Belarussian economy. Indeed economic growth of 10.8% in Belarus in 2004 was largely driven by the growth of exports to Rus-

---

[75]   The report from the meeting of Gazprom's chairman, A. Miller, and Gazprom's deputy chairman, A. Ryazanov, with Belarussian reporters. Gazprom's website, 17.05.2004.

[76]   *IMF country report*, no. 04/141 (May 2004).

[77]   И.А.Михайлова-Станюта, 'Открыта ли Белорусская экономика?', *Белорусский банковский вестник* (Апрель 2003), стр. 46.

sia. The fact that high growth rates result from barter-based exports to Russia implies that Belarussian industry produces more and more goods of mediocre quality, which are impossible to sell for cash elsewhere[78]. Russia so far accepted barter trade because Belarussian goods were cheaper, and also because Russian producers did not have enough spare capacity to produce themselves. However, the situation is changing since the Russian increase in gas prices makes Belarussian goods more expensive, and thus less competitive. Furthermore, on its way to the WTO, Russia aims to restructure and modernize its industry, which implies goods' quality improvement and an increased availability of spare production capacity. This, in turn, would decrease the share of the Russian market currently occupied by Belarussian producers, unless the latter come up with more competitive goods.

It is sometimes argued that the position of Belarus on the Russian market is safe in any event due to the high degree of complementarity of the two countries' productive structures that makes Belarussian goods necessary for Russian producers. But, even if this holds true for some time, such complementarity will weaken as Russia modernizes its economy, and Belarus retains an outdated, energy-ineffective industry. Belarus's ability to modernize its industry prior to losing the Russian market depends, among other factors, on how successfully it integrates into the world economy. Such integration is not going to be easy since the Belarussian economy is highly protectionist and is currently little involved in international markets. Its exports to 'far abroad', i.e., the most effective ones, constitute only 22.4% of all exports. It is often argued that Belarussian economy is the open economy since it is characterised by a high ratio of foreign trade turnover in GDP (105.9% on average during 1991-2001 decade against 40-60% in developed market economies) and a high share of export/import in GDP (50% against 20-30% in developed market economies). However, as I. Mikhaylova-Stanyuta argues, such high figures suggest that the economy is open only if it has a high level of specialisation, which is not the case of Belarus.[79]

---

[78]   By March 1999 inventories constituted 40% of GDP, and by 2003 - 70% of total output. See D. Mario Nuti, *op. cit.*, p. 115.

[79]   For discussion see И. А. Михайлова-Станюта, *op. cit.*, pp. 45-49.

An accelerated production of goods, which are traded on unfavourable terms through barter or unpaid transactions, and a failure to produce goods for which there is market demand, results from the command nature of the Belarussian economy[80]. As our analysis suggests the reforms in Belarus did not fail – they simply have not been tried – and a mixture of a few market measures and non-market administrative controls did not turn the Belarussian economy into a market one. Still, administrative controls, preferred by the government, do not work as intended because some market elements are in place, undermining the system's response to administrative measures. Furthermore, due to its high dependence on foreign trade, the national economy is influenced by economic policies taken by its trade partners. Also, being a member of the Customs Union, Belarus is subject to the additional regulations in the union's framework.

---

[80]   Mario D. Nuti, 'Belarus: a command economy without central planning', *Russian and East European Finance and Trade*, 26:4 (July-August 2000).

2.2   Can Belarus Afford to Pay Increased Prices for Russian Gas Imports?

Gas prices increase is the most obvious, though, as the 2004 gas cut offs suggest, not the most effective way for Russia to exercise its leverage over Belarus, which is 90% dependent on imported gas. Belarus was able to maintain its competitiveness on the Russian market despite its labour productivity that is lower than in Russia because of the republic's secured ability, until recently, to have access to cheap Russian gas.

The reasons for Belarus being able to receive gas cheaply were conditioned by: 1) the Customs Union Agreement according to which all supplies, including gas, are priced in domestic prices of the manufacturing country, 2) the Intergovernmental Agreement, according to which Belarus agreed to privatise its gas transit network and establish the JV to operate its transit routes jointly with Gazprom. Since Belarus failed to fulfil the second part, Russia increased gas prices for Belarus and they ceased to be equal to Russian domestic prices. Importantly, this book argues that, starting from 2004, Belarus will be unable to re-gain its access to cheap Russian gas even if the government does eventually decide to form the JV because gas prices consistently increase and will continue to do so in Russia itself[81].

The book argues that earlier on the Belarussian government had more sources available to keep the Belarussian economy going despite its being largely ineffective. These sources included Russian energy subsidies (essentially the difference between current and pre-crisis gas prices), Russian stabilisation loans and an acceptance of barter trade as means of payments for gas. By these means, during the first quarter of 2003, Belarus was able to make its energy arrears to Russia fall from $314 mln in 2002, to $191 mln (of which $148 mln was for gas)[82]. Much of the reduction was attributable to a $40 mln loan provided by the Russian government following an agreement be-

---

[81]   'Russia agrees to raise prices as part of EU agreement on WTO accession', *European Gas Markets*, 11:05:2 (May 2004), p. 11.
[82]   *Transition Report 2004* (London: EBRD, 2004).

tween Belarus and Gazprom in November 2002 regarding the joint venture. In addition, partial repayment of Beltransgaz's arrears to Gazprom was financed by Belarussian banks acting on the government's orders. As of February 2004, however, Belarus still owed Gazprom $120 mln[83], as well as its debt to independent Russian gas companies, including $41 mln to TransNafta.

Importantly, if barter was widely accepted earlier as a means of payment for gas, the current situation has changed with gas suppliers insisting on cash payments. Prior to 2004 barter was especially widespread in Belarus's gas trade with independent Russian gas suppliers. This was especially true for the Itera company, which was an effective barter trader, and 80% of whose trade with Belarus was made up of barter as late as 2000[84]. However, the share of independent producers supplying Belarus with gas has been consistently decreasing, and since mid-2004 all gas to Belarus is supplied by Gazprom. Therefore the acceptance of barter as the means of payments for gas is likely to decrease, thus further limiting Belarus's financial ability to pay for Russian gas.

From January 2003 Gazprom, witnessing no progress on Beltransgaz JV, increased gas prices for Belarus by some 50% (from $18-22/mcm in 2002 to $30/mcm), which raised import costs by 1.2% of 2002 GDP. Furthermore, starting from mid-2004, Gazprom supplied gas to Belarus at $46.68 – amounting to a further 50% increase. Thus total gas prices increase during 2002-2004 raised import costs for Beltransgaz by 3.2% of 2002 GDP, which is not a negligible figure if the state budget deficit stands at 1.8% and current account deficit at 2.6%.

Since gas prices have increased and payment terms hardened, it is important to assess the financial sources which the Belarussian government currently has at its disposal or can potentially mobilise – such as FDI, domestic savings, revenues from privatisation, foreign loans etc - to cover more expensive gas imports. This book argues that Belarus's ability to pay for the increases is very

---

[83]   Gazprom's data.
[84]   Moreover this share was seen as a substantial improvement compared with 1999.

limited as the national current account has remained in 3% deficit since 2000, even given that gas prices were privileged then and payment terms were lax. Despite a slight increase, privatisation revenues remained low, being only 10% of the 2003 target, and 13% of the 2004 target. The cumulative returns for 1995-2002 constituted only 3% of GDP. Confinement of privatisation to a small-scale and a lack of domestic savings are the most important reasons for low privatisation revenues.

Another possible source of revenues is FDI, but Belarus so far has received only a very modest net flow, with cumulative FDI during 1989-1998 constituting only 1.6% GDP. By the end of 2003 it constituted only 3.4% of governmental needs. Such a low level may be attributed to political and economic factors. Politically, the trend towards greater governmental control over the economy, exemplified by an introduction of 'golden share' rule giving the government a final say in all enterprise's affairs, have led potential investors to worry that the state does not attach much importance to property rights. The rule also enables the government to claim a 'golden share' unless an enterprise is 100% private, irrespective of its shareholdings. Moreover, the 2004 decree extended the potential application of the 'golden share' to any company in which the state at one time held a stake, thus enabling the government to exercise its will in *all companies with the only exception of de novo private firms,* constituting as we pointed out above, only a tiny proportion of the economy. Interestingly, as M. Nuti notes, the 'golden share' right has so far only been exercised on very few occasions, yet it added greatly to uncertainty and thus deterring investors[85].

Economically, the gross overpricing of those Belarussian assets that are available for sale is the major factor discouraging investors and contributing to the low effectiveness of investment in Belarus. For example, the failure to privatise Belarussian petrochemicals via direct sales has been explained by the too high prices initially set by the government and strict requirements on future owners' post-privatisation plans such as maintaining a certain level of em-

---

[85]   D. Mario Nuti, 'The Belarus Economy: Suspended Animation between State and Markets', p. 112.

ployment at the expense of profitability. The story was repeated with the attempt, jointly with Gazprom, to privatise Beltransgaz, which the government valued at $5-6 bln, and Gazprom – at $500-600 mln. Predictably, there were no other bidders apart from Gazprom, partly because no foreign company wanted to clash with the Russian gas monopoly, partly because of the unfavourable political image of Belarus, partly because of the asset overpricing[86]. Thus the book contends that political and economic factors in Belarus reinforce each other in creating a negative investment climate, thus preventing FDI from becoming a reliable and sufficient source of cash inflow to the economy.

Loans constitute another possible means of covering the current account deficit. However, Belarus's access to external financing remains very limited. Thus disagreements over appropriate macroeconomic policies led governmental authorities to withdraw their request for an IMF stand-by program in 2004, repeating the situation in 1998, when the government would not accept a $100 mln post-1998 crisis IMF stabilisation loan for fear that conditions attached to the loan might cause a sharp decrease in living standards, and, consequently, threaten government's survival.

A loan by Belarussian banks is another possible domestic means of financing gas prices increase. But the national banking system is seriously weakened by long-standing governmental orders to support loss-making enterprises, especially in the agricultural sector, contributing towards the erosion of the banks' loan portfolios. Therefore, the national banking system cannot be counted upon as a stable source of finance since it is not rooted in the efficient real sector.

Therefore, as our analysis suggests, the financial resources available to the Belarussian government to cover price increases are neither numerous nor sufficient, with Russia being the only available source of external financing.

---

[86]   The only substantial deal that has taken place so far and contributed to FDI flows of $434 mln was the sale of Slavneft in 2002. *Transition Report 2004* (London: EBRD, 2004).

Thus apart from Russian loans (which are provided because Gazprom is trying to settle the gas conflict peacefully without resorting to a yet another gas war as in 2004), Belarus does not have enough sources to raise additional funds to cover the gas price increases[87].

---

[87]   In August 2004 Russia proposed to extend $175 mln loan to Belarus to help Belarus meet the higher cost of gas imports.

2.3.   Implications of Gas Price Increases for the Sustainability of the Belarussian Economy

This book contends that, by increasing gas prices and insisting on cash payments, Russia in essence forces the Belarussian government to introduce economic reforms that will turn Belarus into a valuable economic partner rather than one having enduring payment difficulties and posing threats to the reliability of Russian gas exports to Europe. Indeed, significantly increased prices will have a devastating snowball effect on the unreformed economy, thus providing a powerful stimulus to introduce structural reforms.

So far the heaviest price pressure was on industrial enterprises since they have long been paying Beltransgaz about twice as much as the latter paid Gazprom. Thus in 2002 Belarussian industrial sector has been buying gas from Beltransgaz at around $40-55/mcm whereas *Belatransgaz* paid only $17-22/mcm to Gazprom and $30-36/mcm to various independent Russian gas suppliers[88]. The difference was retained by the state and contributed toward its budget, to be further used to subsidise goods so that they subsequently enjoyed a competitive price advantage when exported to Russia[89]. As prices further increased for Beltransgaz in 2003, it began to charge the enterprises yet more keeping the pace with prices increases, thus strengthening the price pressure not only on the industrial sector but also on households because industrial enterprises had difficulties putting up with price and required by the government wage increases. The weakened households in turn had difficulties paying their rents and taxes, thus further weakening the state budget and the banking system. Therefore, as prices reached the $46.68 level for Beltransgaz itself, it was forced to reduce its profit margin while selling gas on to industrial enterprises. Thus the difference between the enterprises' and Beltransgaz's expenditure began to shrink.

---

[88]   Itera, TransNafta and SIBUR.
[89]   Interview with Itera's president Makarov, *Gas Matters* (November 2002), pp. 12-16. See also 'Gazprom Not Happy with Belarus', *RFE/RL,* 12 November 2002.

This book evaluates how this difference changed since 2002, when Belarus began to receive gas at Russian domestic prices, and through the successive years as Gazprom increased gas prices for Belarus in response to Belarus's refusal to establish a JV. We based our evaluations on the data presented in Table 2.2 and summarized our findings in Table 2.3 below.

We found that the difference between the industrial and Beltransgaz's expenditures declined from $503 to $478 mln, and is projected to decline still to the $445.5 mln level in 2005. We argue that since this difference earlier served as an important source of budgetary subsidies, its decrease will undermine the very basis of the Belarussian 'welfare state'.

Table 2.2 Dynamics of Gas Prices, Transit Fees and Volumes of Gas Supplied to Belarus by Gazprom and the Independent Russian Gas Companies*

	1999	2000	2001	2002	2003	2004	2005
Volumes, bcm, supplied by:							
1.Gazprom	12.2	10.8	16.2	10.2	10.2	10.2	19.1
2. Independents							
- Itera	4.3	5.8	5.1	5.9	6.3	6.4	---
Transit fees, $ per 1 mcm/100 km, charged by:							
1. Belarus							
- via Beltransgaz-owned network;	0.53	0.53	0.53	0.53	0.53	0.75	0.75
- via Yamal-Europe;	0.46	0.46	0.46	0.46	0.46	0.46	0.46
2. Ukraine	1.09	1.09	1.09	1.09	1.09	1.09	
3. Poland						2.74	2.5
Gas prices, $ per mcm:							
1. Russian domestic prices, 9th zone	11	13.3	16	19	23	28	40
2. Border price for Europe					100-125	150	
3. Price paid by Beltransgaz to:							
- Gazprom	30	30	30	17-22	30	46.68	46.68
- Itera				30-40	46.68	46.68	---
4. Price paid by Belarussian enterprises to Gazprom		40-55	40-55	40-55	60-65	70	

Source: various issues of *Cedigaz, Gas Matters, European Gas Markets*

Table 2.3: Dynamics of Gas Imports Expenditures

	2002	2003	2004	2005
Beltransgaz expenditures ($ mln) on gas supplied by:				
Gazprom	224.4	306.0	476.1	891.5
Itera and the other independents	295.2	387.4	480.1	---
A. Total	519.6	693.4	957.2	891.5
Volumes, bcm	18.4	18.5[90]	20.5	19.1[91]
B. Industrial enterprises expenditures	1023.0	1184.0	1435.0	1337.0
The difference between Beltransgaz's and industrial enterprises' gas expenditures (B-A)				
- absolute, $ mln	503.4	490.6	478.0	445.5
- % of GDP	3.7	2.9	2.2	na
- % of subsidies	19.5	15.6	11.9	na

Source: calculated from Table 2.2 and form various issues of *Cedigaz, Gas Matters, European Gas Markets*

Indeed, if in 2002, when gas was shipped to Belarus on the most favourable terms ever, the difference amounted towards 19.5% of all state subsidies, this

---

[90] There is some uncertainty concerning volumes of gas supplied to Belarus by Gazprom in 2003. According to official Gazprom's data it supplied 18.1 bcm, and the independents supplied remaining 6.2 bcm. However, it is unclear what Belarus would have been doing with such large volumes. Indeed, other sources such as *European Gas Markets*, 10:10:1 (October, 2003) and *Western Gas Intelligence*, 15:2 (January 2004) evaluate Gazprom's supplies as some 10.3 bcm thus suggesting that Gazprom may have included in the reported 18.1 the volumes supplied by the independents.

[91] For all years up to 2005 all volumes actually delivered, for the year 2005 – volumes contracted.

proportion declined to the 11.9% level in 2004 whereas the share of subsidies in GDP remained unchanged at the 18.9% level. As long as GDP grows fast enough this decline could be covered by incremental growth of GDP. However, as we pointed out earlier, since the growth of the Belarussian GDP is mainly driven by Belarus's exports to Russia, its prospects are questionable in view of decreasing price competitiveness of Belarussian goods earlier provided by low gas prices.

Having concluded that even current gas prices, i.e. $46.68/mcm, are already too heavy a burden for the Belarussian economy, the book further contends that these prices are the very minimum that Belarus should expect to pay because they are, in effect, *already equal to Russian domestic gas prices plus the transportation cost*, and Russia cannot and will not sell gas to Belarus at lower prices than those it charges its domestic consumers. Moreover, we contend that gas prices for Belarus will further increase because Russian non-residential gas prices are increasing and will continue to do so in line with Russia's economic and energy strategy and also in line with WTO requirements. Indeed, in February 2005, Gazprom's Deputy Chairman made a statement that Gazprom will require an increase of the current - $ 46.68/mcm - price as early as 2006. Therefore even if Belarus eventually agrees on a JV establishment, there will be no return to low gas prices[92].

Thus the Belarussian government's resources to pay for gas and sustain a so called 'welfare state' will be further depleted as it will have to cut off financing such elements as wages in excess of productivity, large subsidies, as well as high pensions. This enables us to argue that the pivotal factor that previously sustained the Belarussian economy and made its non-reforming feasible was cheap Russian gas. We agree here with M. Nuti that the Belarussian economy could carry on being unreformed if it had continued access to cheap Russian energy[93]. If such an access is denied then the main source of Belarussian welfare disintegrates.

---

[92]   'Газпром' стимулирует в Беларуси экономические реформы?' *Экономика и Бизнес,* Март 2005.

[93]   M. Nuti, *op. cit.*, pp. 113,119.

Despite Gazprom's threats to suspend gas deliveries as early as 2002 if Belarus did not accelerate the establishment of a JV, Belarus did not even undertake any gas–saving measures, let alone restructuring of its highly gas-intense industry, before Russia cut off gas in February 2004. Only after that happened did the government introduce its Energy Saving Programme[94]. If properly implemented even marginal economising measures could reduce industrial gas consumption up to 25%. However, since Belarussian industry is 90% gas-fired only deep structural reforms could reduce gas consumption radically.

This book contends that there is an important trade-off between the Belarussian leadership's ability and willingness to reform the economy, and there are serious impediments in the way– both political and economic. While fearing to embark on reforms because they would provoke political instability due to the initial deterioration of living standards, inevitable gas prices increase would also lead to a worsening of the economic situation unless gas conservation measures are adopted.

Thus Belarus no longer has a choice between reforming and non-reforming because in the absence of Russian economic patronage non-reforming is as dangerous for the regime's stability as reforming. If successful the reforms would lead to gas conservation and improved competitiveness of the national economy, thus decreasing the republic's vulnerability to gas supply shocks. At the same time, continuous postponement of reforms, as the major source for making non-reforming viable deteriorates, can trigger even more regime instability than the reforms themselves. Furthermore the failure to introduce reforms earlier limits government's choice of reforming policies now. Indeed, had the reforms begun when gas prices were low, they would have cushioned current welfare deterioration provoked by increased post-2004 gas prices, thus posing less of a threat to the regime and providing more time for reforms.

However, the book argues that even if the leadership were willing to embark on reforms, it would be difficult to implement them while maintaining its undemocratic nature. Though, as C. Lawson correctly notes, the implementation

---

[94]   http://www.government.by/en/eng_solution341.html

of such reforms requires a sustained political will that is 'very hard to exercise in a democratic system'[95], we argue that a democratic system is still more conducive to market reforms than an authoritarian one. Indeed, market reforms are impossible to implement voluntarily, top down, since administrative controls are less effective than price controls. Furthermore, since Belarus has not undertaken reforming for nearly a decade, its interest groups (such as ex-communists, ex-state-managers, state trade unions) have had more time to perfect their rent-seeking and 'blocking techniques' thus making the implementation of reforms all the more difficult.

Belarus's ability to reform is further limited by its semi-isolation. The stance adopted by its government seriously hampered Belarus's chances of joining the WTO, one of the most effective ways to accelerate restructuring the national economy[96]. Formally only countries with market economies can apply to join the WTO. However, since there is no strict criteria to measure whether an aspiring country is a market economy or not, as long as both the EU and the US *tend to see market economy and democracy as a package*, the undemocratic nature of Belarus diminishes its prospects becoming a WTO member. But since Russia is interested in Belarus joining WTO since it could stimulate domestic economic reforms, Russia can play an important role in facilitating Belarus's entry by exercising its political leverage over the incumbent leadership.

---

[95]   Colin W. Lawson, *op. cit.*, pp. 176-194.
[96]   П. Никитенко (ред.) *Всемирная торговая организация. Беларусь на пути в ВТО* (Минск: ИООО Право и экономика, 2002), стр. 54-59, 106-107.

3   Russian Gas to Europe: An Analysis of Existing and Projected Export Routes

This chapter explores how Russia's gas export strategy towards Europe is affected by 1) existing gas routes going through Belarus, 2) routes passing through Ukraine, 3) the new offshore projected route by-passing all CIS transit countries. The book explores the rationale behind the projected route selection and its implications for the usage of existing routes. It explores how reliable the existing routes are and analyses how changes in the reliability of one route affect the degree of reliability of the others.

3.1   Russia and the Gas Supply of Europe: Dependence, Diversification and Export Routes

Europe, aiming to reduce its reliance on oil, faces a choice as to where its future incremental gas imports will come from. In view of the recent EU statement that there is not, and cannot be, any limitation on Russian energy deliveries to Europe, Russia stands a good chance of increasing its presence in the European gas market.

Russia is interested in expanding its European exports since Gazprom provides around 25% of Russian federal budget revenues. In turn Gazprom itself is heavily dependent on export earnings for profitability and loan guarantees, particularly since it was only in 2004 that domestic gas sales became profitable[97]. Thus an 'emphasising value' approach, being the most important change in Gazprom's post-USSR export strategy, explains Gazprom's desire to keep and expand its presence in Europe[98]. Here it faces two challenges: to keep and increase its market share among 'old' Europe's countries and not to lose a traditionally high share of new member-states' gas markets (see Table 3.1).

The first task is the less difficult one, since western European countries themselves want to increase their gas imports from Russia. Eastern European countries, by contrast, want to diversify their gas imports being currently almost totally dependent on Russian imports[99].

Thus Poland, the Czech Republic and Hungary have already attempted to bring non-Russian gas supplies to their markets. However, Russian gas being the cheapest option, diversification is costly and financially difficult even for most economically advanced eastern European countries[100]. This suggests

---

[97]   Jonathan Stern, *op. cit.*, p. 141.

[98]   'Gazprom reaffirms plan to export 153 bcm to Europe in 2004', *European Gas Markets* (31 March, 2004).

[99]   See *BP Statistical Review of World Energy*.

[100]   'EU accession – ten new countries, and a few new problems', *Gas Matters* (May 2004).

that their diversification attempts are driven by concerns about possible political consequences of potential physical dependence on Russian gas rather than by economic considerations. Also, such attempts reflect a rejection of historical memory of Soviet dominance. However, even if eastern Europe is to succeed in diversification, it is rightly argued that its *physical* diversification, as opposed to *contractual,* may well be an illusion, and the countries will still receive Russian gas even supplied by non-Russian contractors. Nonetheless, even if the diversification is of a symbolic nature, it does suggest that eastern European countries aspire to continue contractual diversification of their gas imports in the future[101]. Thus in spite of the fact that Gazprom's exports to eastern European countries remain high, they fell slightly in 2004[102].

---

[101]   Jonathan Stern, *op. cit.,* p. 170.

[102]   During the first two months of 2004 against the same period in the previous year (8 bcm and 7.1 bcm of gas, respectively). See A.Ryazanov and A.Medvedev' press-conference, Gazprom website, 2.03.04.

Table 3.1: Russia's Gas Exports to Europe in 2004 and Their Share in Domestic Consumption (top 10 importers), bcm

	Russia's gas exports, bcm	Domestic consumption, bcm	Share of Russian gas in domestic consumption, %[103]
Total exports	149.0		
Germany	40.8	94.5	43.2
Italy	21.5	80.7	26.6
Turkey	14.5	22.2	65.3
France	14.0	47.9	29.2
Hungary	9.2	13.7	67.2
Slovakia	7.7	7.7	100.0
Czech Republic	6.8	9.0	75.6
Poland	6.3	13.1	48.1
Austria	6.0	9.5	63.2
Finland	4.9	4.9	100.0

Source: Calculated from Gazprom data and Cedigaz

As for 'old' Europe, its gas supply is well-diversified, not in the least because there was an unwritten Cold War understanding that Russian gas imports should not exceed 1/3 of western European needs.

Continuing liberalisation of the European gas market, followed by prospects of short-term trading can weaken Gazprom's financial position being detrimental to the security Gazprom gains from planning. But as long as Gazprom holds such a substantial share of the European market as 30%, liberalisation would not weaken its position substantially. An awareness of this fact makes Gazprom even more willing to increase its European presence. According to Gaz-

---

[103]   These figures are approximations

prom's understanding such a position can only be maintained by selling gas as far from its production source and as near to its end-user as possible.

In pursuing this goal Gazprom devised a 'downstream' strategy of forming joint ventures to prevent the sole importing national company from getting a large share of Gazprom's sales profit[104]. Gazprom's export affiliate, Gazexport, was active in this respect in eastern European countries having successfully entered national distribution systems in Romania, Hungary and Slovakia, though failing to do so in the Czech Republic[105]. The main purpose of Gazprom's strategy of establishing joint ventures in Europe is to market Russian gas and secure higher profits. There is a different rationale behind its attempts to create joint ventures in CIS transit countries. In Belarus and Ukraine its aim is to achieve control over networks to ensure uninterrupted transit.

To further maintain its 'closer to consumer' strategy Gazprom needs reliable and well-diversified gas evacuation routes from Russia to Europe (see map below for gas export routes, Fig. 3.1.

---

[104]   Jonathan Stern, *op. cit.*, p. 162.
[105]   *Ibid.*, p. 168-169. The Czech Republic proved to be the most decisive in its intention, being the only 'major Russian gas buyer' that rejected Gazprom's offer to form a joint marketing company. See A.Miller and A.Ryazanov' press-conference, Gazprom's website, 17.05.2004.

Fig. 3.1 Russia's Gas Export Routes

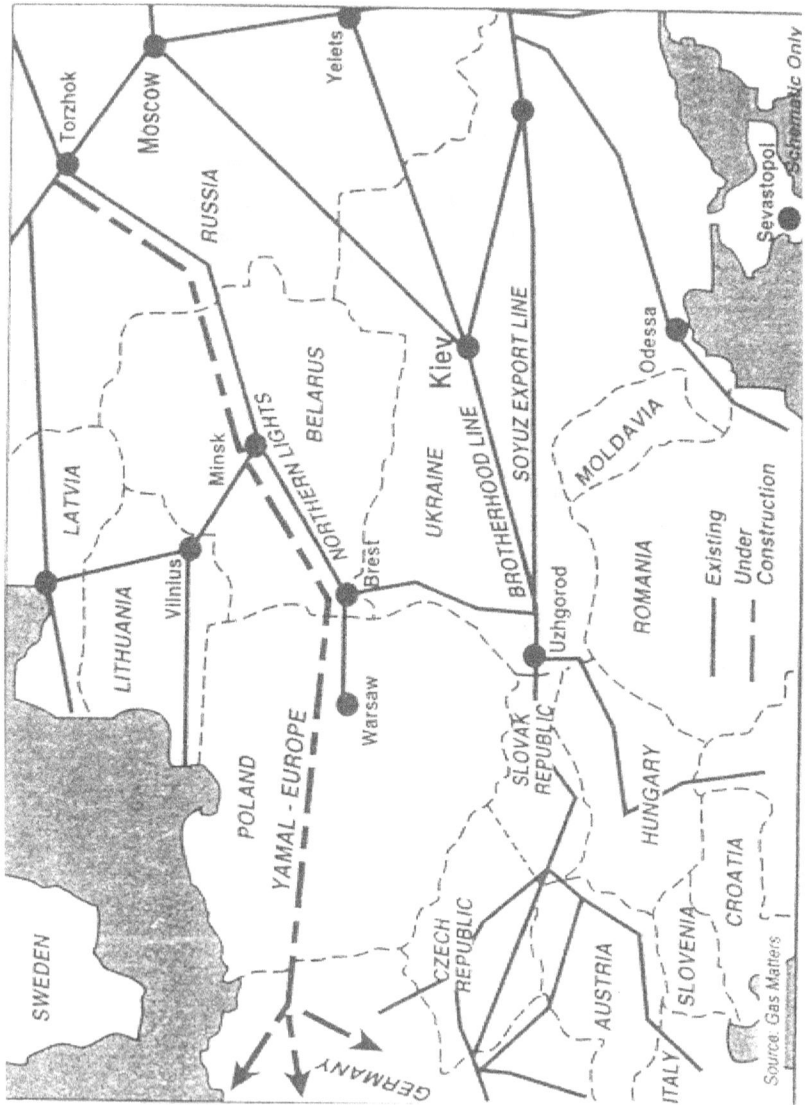

Source: Jonathan Stern, *Russian and Soviet Gas*.

Thus Gazprom undertook significant investment in building new pipeline systems such as Yamal-Europe, following transit problems with Ukraine, and Blue Stream, following transit problems with Bulgaria. Gazprom also devoted much effort to promoting the Ukrainian bypass via Slovakia, though only to abandon it thereafter. Most recently Gazprom considered building an offshore line under the Baltic Sea, following transit problems with Belarus, which would finally avoid *all CIS and eastern European countries*[106].

However, having more pipelines running through transit countries does not necessarily imply a higher reliability of gas flows. Indeed, while the growth of regional gas pipeline grids links many nations commercially, having important geopolitical and security implications, it also raises the vulnerability of grid members to political hostility from each other. Gas disputes in the CIS, the first being between Gazprom and Ukraine, and the second - between Gazprom and Belarus, have demonstrated this. At the same time, gas pipeline construction is highly capital intensive and the inflexible supply system makes it difficult to recover from the consequences of unfavourable investments.

---

[106]   Reportedly, Gazprom's strategic target is to limit transit through any one country to 30-40% of total export volumes.

3.2   The Gas Transit Routes: Belarus and Ukraine

Before exploring the options available for Gazprom to construct new gas evacuation routes, this book analyses the existing routes passing through Belarus and Ukraine. We start our analysis in Ukraine because the growing unreliability of Ukrainian transit routes was one of the major reasons that determined Gazprom to invest in the Yamal-Europe pipeline via Belarus in the early 90s. This book employs a comparative approach, analysing the Belarussian routes and contrasting them with the Ukrainian ones, because changes in the reliability of the Ukrainian transit system influence the relative importance of the Belarussian one, and vice versa. An actual or anticipated change in the reliability of Ukrainian routes increases Gazprom's willingness to seek a compromise with Belarus. For example, between 2002-2004 – a period marked by some improvement in relations between Gazprom and Ukraine – Gazprom started to perceive the Belarussian routes as somewhat less important. However, in view of political uncertainty in the aftermath of December 2004 Ukrainian presidential elections the Belarussian routes regained their significance.

Country Profile: Ukraine, the Transit Routes.

As much as 80%[107] of Russia's gas exports to Europe[108] flow via the Ukrainian gas transit network, which was built in the Soviet era and became the property of Ukraine at independence (see map below, Fig. 3.2). The network is old and is in a state of great disrepair. Its refurbishment, while possible, requires substantial investment, which, however, is less expensive than building a new pipeline[109]. But the Ukrainian state-owned gas company, Naftogaz Ukrainy, does not have sufficient funds to refurbish the network, whereas Gazprom does not have a desire to invest into assets over which it exercises no control.

---

[107]   Reduced from 90% with the Yamal and Blue Stream pipelines opening.

[108]   In 2004, Russia (i.e., Gazprom, since it holds an exports monopoly) exported to Europe 140.5 bcm of Russian gas and around 8 bcm of Central Asian gas.

[109]   Interestingly, under Viakhirev's chairmanship, Gazprom made the opposite statements - that it would be financially preferable to build a Polish-Slovakian bypass rather than upgrade the existing Ukrainian network.

Fig. 3.2 The Ukrainian Gas Transit Network

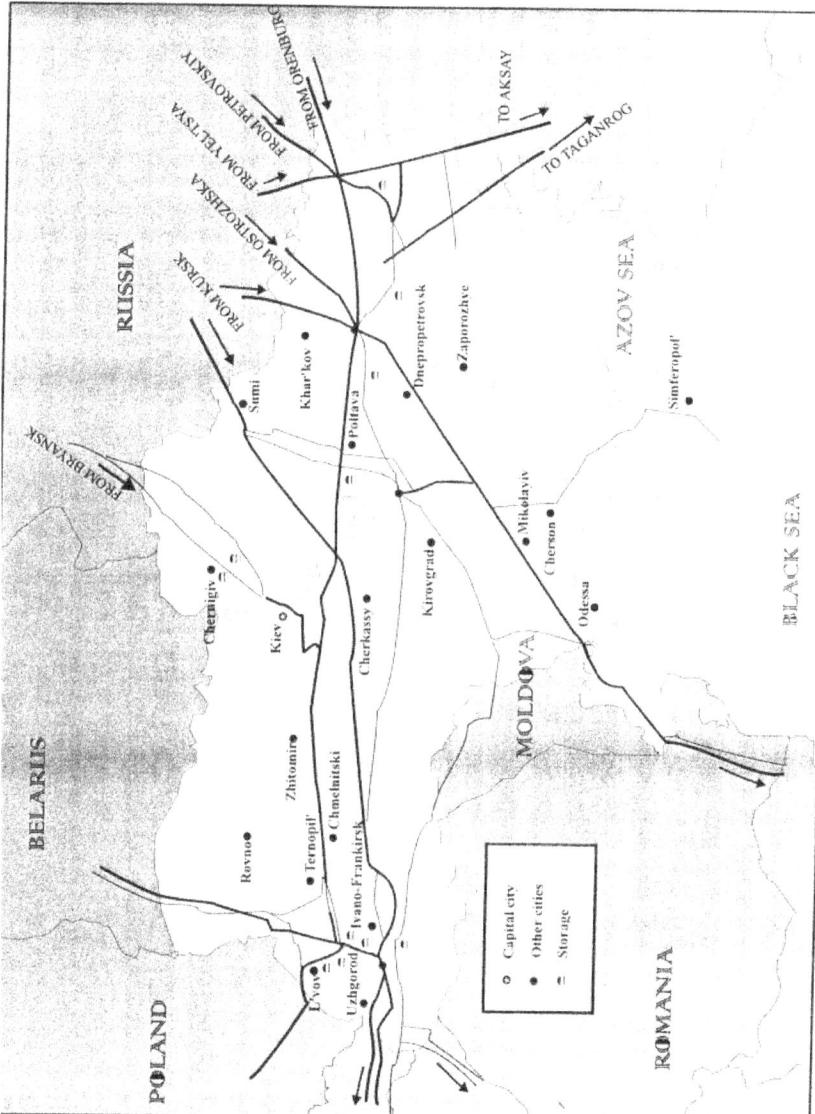

Source: *European Gas Markets,* 4:04:1, 2001.

The absence, over a decade, of sufficient investment in pipeline maintenance, the enduring nature of unauthorized gas offtakes, and the lack of prospects to improve the situation, led Gazprom in 2000 to state its aim to reduce gas transit via Ukraine by 2/3 in 6-8 years by rerouting the gas northwards – via Belarus and Poland – using Yamal-Europe and southwards – across the Black Sea to Turkey – using Blue Stream[110].

Ukraine, similarly to Belarus, is heavily dependent on gas imports as 80% of its gas requirements either originate in Russia or travel via the Gazprom-owned network from Turkmenistan[111]. Ukraine's domestic production constitutes around 17 bcm and accounts for only 20% of the country's gas needs. Ironically, until the mid-70s, Ukraine was a major gas-producing region of the USSR and it was Ukrainian gas that was exported to Europe. However, after Ukrainian production peaked it was abandoned, and the USSR accelerated its investments in Russian Western Siberia[112]. Nonetheless, Ukraine, in sustaining attempts to reduce its dependence on Russian gas, actively promotes new gas field discoveries on its territory[113]. Despite that, Ukraine will remain heavily dependent on gas imports for the foreseeable future.

Following independence, Ukraine's terms of gas trade with Russia deteriorated as gas prices increased sharply. An inefficient gas user, Ukraine had accumulated large debts to Russia (particularly to Gazprom and Itera) and Turkmenistan. Since the volumes of gas delivered via Ukraine are large, Ukraine's non-payment meant that Gazprom was losing significant sums. Ukraine also illegally siphoned Russian gas out of transit pipelines for further resale and domestic use on a large scale – the task greatly simplified by the separation of transit and gas purchase. However, because around 120 bcm of Gazprom's gas exports to Europe transit via Ukraine, it has a considerable leverage vis-à-vis Gazprom regarding gas volumes, prices, transit fees, and

---

[110]   'The Russian Revolution', *European Gas Markets*, 7:04:1 (April 2000), pp. 6-7.
[111]   59% of Ukrainian gas imports originate in Turkmenistan.
[112]   Caspian sea and Ukraine's quest for energy autonomy. *Geopolitics of Energy* (October 1998).
[113]   'Ukraine boosts its domestic gas production', *European Gas Markets*, 8:8:1 (August 2001), p. 12.

debt treatment[114], thus balancing its gas supply vulnerability. Indeed, Gazprom (unlike Turkmenistan), being cautious about its reputation as a secure supplier of gas to Europe, never cut deliveries via Ukraine.

Gazprom's exasperation with illegal Ukrainian siphoning and increasing gas debt gave momentum to the decision to build an alternative route, the Yamal-Europe pipeline, via Belarus and Poland. However, even if both planned strings are built - which is unlikely given the unresolved Russian-Belarussian gas dispute – the pipeline would still not be sufficient to compensate Gazprom for not-using the Ukrainian network since Yamal's onward exports to Euro-pean countries other than Germany are limited[115]. Therefore Gazprom, being interested in increasing the reliability of the Ukrainian routes, is trying hard to gain some control over the Ukrainian gas transit network.

In pursuing that goal Gazprom imposed complex and stringent arrangements on Russian gas supply to Ukraine in 2001 intended to stop illegal gas siphon-ing, non-payments and resale. A clear sign of Russia's determination to in-crease its influence over the Ukrainian gas industry followed with the appoint-ment of former Russian prime minister, and the head of the Soviet Gas minis-try in the 80s, Viktor Chernomyrdin, as ambassador to Ukraine. This appoint-ment was the most promising way for Gazprom to achieve its aims because of Chernomyrdin's personal contacts, which matter more than paper agreements in post-Soviet states.

Ukraine's high dependence on, and especially its indebtedness for, gas en-abled Gazprom to press it into swapping its transit network and storage facili-ties for debt. However, whereas in 1994 the Ukrainian government had initially agreed to exchange shares in its transit network, Ukrtransgaz, for gas debt, the Ukrainian legislature refused to approve the deal perceiving it as being

---

114   In 1997 Ukraine relieved 30 bcm of gas from Gazrpom in lieu of transit fees – roughly equal to total Gazprom's deliveries to Ukraine and half of the total coun-try's needs, 77 bcm.
115   *Western Gas Intelligence (*September 1999), p. 3.

detrimental to the country's sovereignty[116]. Such an outcome was possible because Ukraine, unlike Belarus, had a strong opposition represented in the national legislature and highly concerned about Russian influence in a sector as
strategic as energy.

Thus as late as December 2000, there was still disagreement over the gas
network privatization bill, with the then Ukrainian prime minister Yuschenko
talking against privatisation on the grounds of the low price that would be ob-
tained for the network because of its 'parlous state of repair'[117]. Other Ukrainian politicians were supportive of privatisation perceiving it as the only way to
solve the gas debt problem and finance the network's maintenance[118]. The option of lending the network to Gazprom for an agreed period provoked objections that Gazprom would 'squeeze' what it could out of the grid without maintaining and developing it. This reaction was quite surprising since it is contrary
to Gazprom's own long-term interests to let the Ukrainian network deteriorate.

Thus, as J. Stern argues, Ukraine's 'political sensitivity' towards Russian
presence was at the heart of its continuous inability to achieve a compromise
commercial solution in which Gazprom would get partial ownership over
Ukrainian network and storage[119]. By the late 90s, however, both countries realised the necessity of establishing some commercial framework, which would
allow to shift gas issues from the political to the economic dimension. An important step, which marked a serious improvement in relations, was the full
and final settlement of Ukraine's debt for natural gas deliveries between 1997
and 2000 stipulated by the agreement signed between Gazprom and the
Ukrainian government in August 2004[120].

---

[116]   'Caspian sea and Ukrainian quest for energy autonomy', *Geopolitics of Energy*
(October 1998).
[117]   21% of all pipelines are past their amortisation period or only have short-term anti-
corrosion protection. See Privatisation is the only way to restore Ukrainian pipeline.
European Gas Markets, 11:12:2 (December 2004).
[118]   *Ibid.*
[119]   Jonathan Stern, *op. cit.*, p. 158.
[120]   Gazprom's website, August 2004.

This agreement gave a fresh start to what had been very strained relations and the sides came up with the idea of establishing an international consortium to manage, operate, and maintain the Ukrainian network[121]. The Russian president has welcomed the idea calling it 'an important step' towards a long-term solution[122]. The consortium envisaged the construction of a new pipeline, which would connect Ukraine's east and west and enable Gazprom to increase its exports via Ukraine by 12-13%. Furthermore, Ukraine agreed to make available to Gazprom an undisclosed amount of underground gas storage capacity on preferable terms to allow the company to maintain steadier year-round production levels[123]. These agreements demonstrated that Gazprom and Naftogaz Ukrainy were entering a new, friendlier, phase of their relationship.

However, prior to the Ukrainian presidential elections in late December 2004, Gazprom and Naftogaz Ukrainy had not decided on the details of participation in the consortium. The major obstacle was Ukraine's insistence on the third-party participation, preferably by a foreign oil/gas company. Whereas initially Rurhgas, ENI and Wintershall expressed some interest in participation[124], there was not really any keen interest on their part - partly because the Ukrainian network requires massive investment while the property rights are highly uncertain, and also because of their unwillingness to clash with Gazprom which clearly preferred to remain the sole foreign partner in the consortium. Indeed, when in late 2002 the intergovernmental agreement between Russia and Ukraine was eventually signed, crucially, no third party was present[125]. Therefore, the scheme is unlikely to succeed given the Ukraine's firm stance to avoid a consortium in which Gazprom is the only foreign party.

---

[121]   *Cedigaz* (August 2004).
[122]   *Cedigaz* (October 2002).
[123]   *Western Gas Intelligence* (June 2002).
[124]   'Ruhrgas to join consortium for Ukraine's gas transit system', *European Gas Markets,* 10:01:1 (January 2003), p. 13. See also *Cedigaz* (February 2003).
[125]   *Western Gas Intelligence* (October 2002). See also 'Russia / Ukraine transit deal signed, but without Ruhrgas', *European Gas Markets,* 9:10:1 (October 2002), p. 8.

The results of the presidential election in Ukraine are likely to strengthen further this attitude because the new leadership will presumably be careful to create a commercial relationship with Gazprom that will provide for a more 'politically acceptable framework' of economic and energy 'interdependence' between Russia and the 'new' Ukraine, making the bilateral Gazprom-Ukrtransgaz consortium less likely. At the same time there are better prospects for establishing a *truly international* consortium, i.e., one in which more foreign companies would participate together with Gazprom, especially if the EU lends its support to such an option.

If this does turn out to be the case Gazprom will be forced to give up its idea of being the sole foreign partner in the Ukrainian network. Indeed Gazprom is not in a position now to raise strong objections towards third-party participation because if the consortium is not established, either with or without a third party, there will be no other way left to attract funds to refurbish the ageing Ukrainian network to keep it in operational condition. Moreover, Gazprom is not interested in a direct confrontation with Ukraine, which still handles 80% of its gas exports to Europe, especially in view of unresolved gas issues with Belarus[126].

Country Profile: Belarus, the Transit Routes.

Belarus, unlike Ukraine, is the country which Gazprom considered its most reliable transit route. This perception owed much to the ongoing process of Russian-Belarussian integration. This confidence, compared with growing uncertainty surrounding the Ukrainian routes, led Gazprom to invest in the new gas export route – Yamal-Europe pipeline – passing through Belarus and thus turning it into an important gas transit country.

However, by 2000, as the integration processes began to slow down, signs appeared that Belarus might cause transit difficulties in the future similar to those posed by Ukraine. Anticipating these problems Gazprom tried to settle Belarussian gas debt and secure some control over the Belarussian routes by

---

[126]   A. Miller and A. Ryazanov' press-conference. Gazprom's website, 17.05.2004.

establishing the JV to operate them jointly with Belarus. Although it initially agreed, Belarus later reneged on the agreement. Trying to force Belarus either to honour an agreement or pay higher prices for gas, Gazprom cut off supplies via the Belarus-owned Northern Lights pipeline, and when Belarus undertook unauthorized gas offtakes from the Yamal-Europe line in February 2004, Gazprom cut off supplies via this line too, thus cutting *all* gas flows via Belarus.

By mid-2004 Gazprom had labelled the Belarussian routes as the 'least secure' for Russia's European exports. The rapid deterioration of their reliability, considering the significant investment sunk into them, is very unfortunate for Gazprom since it seriously limits its ability to ensure uninterrupted gas exports to Europe, particularly in view of the uncertain prospects of Ukrainian transit routes.

This book examines whether Gazprom can use other networks to insure uninterrupted exports to Europe and analyses the existing routes' expansion and new routes' construction plans.

Existing Routes Via Belarus

The Northern Lights pipeline system running through Belarus and connecting Russian Torzhok with Poland was built during the Soviet era and is used for gas exports to Poland, Belarus and the Baltics. Like the Ukrainian network, it became Belarus's property at independence. Being newer, it is in a better technical condition than the Ukraine's, although it still requires substantial investment to prevent it from deteriorating.

The Yamal-Europe (Yamal-1) pipeline was constructed by Gazprom in post-Soviet times to expand gas exports to Europe and diversify export routes. After the break-up of the USSR the major ideological limitation on expanding Soviet gas exports to Europe disappeared. Thus, already in 1992, Gazprom, eager to expand its European exports, began to work on the idea of establishing a new export corridor to European markets. A year later Gazprom finalized its plans for Yamal-Europe - a system of trunklines running from the Yamal Peninsula to Torzhok from where the routes followed the Northern Lights system, going through Belarus to Poland, and into Germany. This line was meant to enable Gazprom to integrate further into the European gas systems and allow it to increase the flexibility of gas export routes.

Export route diversification was another reason behind Yamal-Europe's construction. Gazprom's growing exasperation with Ukraine's non-payments, gas siphoning, and its inability to invest in the network maintenance, contributed greatly to Gazprom's determination to accelerate construction of a new transit route avoiding Ukraine.

Yamal's construction was based on intergovernmental agreements with Belarus and Poland and was carried out with the co-operation of Polish and German partners. The project was very costly, requiring about $4 bln, and Russia built and paid for the vast majority of the Belarussian section of the pipeline. The first deliveries from Yamal-Europe began in the autumn 1999 and the full completion of Yamal's first string was planned for mid-2005. However, so far 4 compressor stations remain to be added in Belarus to bring the line to its full capacity, 33 bcm per year[127].

By the mid-90s, however, it became clear that European consumers would not need long-term contract gas in such large quantities as early as 2000, and that Yamal at its designed capacity of 33 bcm would not be filled before 2005. Expansion of export capacity ceased to be an urgent task for Gazprom and the work on Yamal line construction slowed down. Moreover, growing difficulties in relations with Belarus culminated in the 2004 gas crisis, which added to the 'evaporation of the urgency' of Yamal's construction. Consequently Gazprom was not willing to accelerate construction of all the remaining compressors, arguing that the first string in its current capacity was fully sufficient for Gazprom's exports to Europe.

---

[127]   Now it is likely that only 2 compressors, instead of the planned 4, will be added, which would provide up to 30 bcm.

Projected Routes Via Belarus:

Belarus's apparent unreliability as a transit country was one of the major reasons behind Gazprom's decision to abandon construction of *the second string of Yamal-Europe (Yamal-2)*. As Gazprom's chairman put it, the company is 'in no hurry' and 'can make up its mind' about Yamal-2 in some 5-6 years[128]. Indeed, if Gazprom were intending to build the second string, it would have brought the first string to full capacity first, but it has not done so arguing that the existing capacity is fully sufficient.

This suggests that Gazprom is now averse of any further investments in transit routes passing via Belarus, be it Beltransgaz's line or Yamal-Europe expansion. Thus Belarus reneging on the JV irreversibly lost potential additional transit revenues that Yamal's expansion would have brought. Furthermore, even if the JV is eventually established, Gazprom will be unwilling to invest in Yamal's expansion, as opposed to its maintenance, because Belarus would still retain some leverage over the pipeline, however diminished, by having the pipeline on its territory. This book argues that Gazprom will probably only consider Yamal's expansion again if Belarus becomes a part of the Russian Federation. But, as argued earlier, there are very few grounds for such a scenario.

The growing unreliability of Ukrainian routes in the late 90s made Gazprom seriously consider the construction of *a major route bypassing Ukraine*. The first phase of this pipeline was envisaged to diverge from the existing first string of Yamal-Europe and to go from Belarus to Poland and then turn southwest past Lublin and from there towards the Slovak-Ukrainian border near Uzhgorod, thereby bypassing the Ukrainian network[129]. The project, however, did not proceed further[130] despite several companies such as ENI, Gaz de France, Ruhrgas and Wintershall, Wingaz and Snam giving their preliminary

---

[128]   The press-conference of Gazprom's deputy chairman, A. Ryazanov, and Gazexport's director general, A. Medvedev, Gazprom's website, 2.03.2004.

[129]   'Russia ups the ante in Yamal re-routing row', *European Gas Markets*, 7.08 (August 2000), p.12. Also see 'Yamal-2', *European Gas Markets*, 8:6:1 (June 2001).

[130]   *Cedigaz* (February 2002).

consent to participate in the international building consortium together with Gazprom[131].

The major reason why the bypass was abandoned is that given its relatively high cost - $1 bln – it would merely re-route existing exports and therefore neither add to Russian export capacity nor increase transit security[132]. Another reason was that Gazprom had not faced the issue of capacity allocation among the partners of the consortium. Following the gas conflict with Belarus Gazprom is especially unwilling to reconsider the route now, even if recent changes in the Ukrainian political system make Ukraine's routes uncertain once again. Indeed, as a line that diverges from Yamal-Europe, the bypass, if built, would only make Gazprom even more dependent on Belarus. Instead, Gazprom is more interested to focus its investment on maximizing the capacity of existing export routes, and when the time comes to commit to a new route, it would opt for an offshore export route that would altogether bypass CIS and east European countries.

---

[131]   'International consortium to spearhead southward diversion of Yamal-Europe', *European Gas Markets*, 7:10:1 (October 2000).

[132]   'Gazprom close to abandoning Yamal 2, its Ukranian bypass', European Gas Markets (January 2002).

3.3   The Offshore Export Route: Is the North European Gas Pipeline an Alternative to the Belarussian Transit Routes?

Gazprom stated that bringing up the first string of Yamal-Europe to full capacity, completing the Blue Stream and expanding the Ukrainian gas network is sufficient to cover forecast European gas demand. However, disappointed about Belarus's and uncertain about Ukraine's routes reliability, Gazprom, in an attempt to avoid any future dependence on transit countries, came seriously to consider the offshore export route option, the North European gas pipeline[133], the NEGP, which aims to transport Russian gas from the Petersburg region to northern Germany, under the Gulf of Finland and the Baltic Sea, with possible routes to the UK, Sweden and Finland (see map below, Fig.3.3). We argue that Gazprom is *forced* rather than willing to consider the offshore route. Gazprom perceives the new export route as more urgent because the problems with Belarus made the monopoly indefinitely postpone Yamal-2's construction which meant to pass via Belarus and cover further incremental increases in European gas demand. This book disagrees with the opinion that Gazprom interrupted the gas supply via Yamal-Europe in February 2004 with the sole purpose of demonstrating to European gas importers the benefits of the direct route so as to attract the additional funds necessary for its construction[134]. Whereas we agree that the EU commission became more supportive of the NEGP after the interruption, it also supported the offshore pipeline before the gas crisis between Belarus and Gazprom. Indeed, as early as November 2002, the NEGP, as well as the Yamal-Europe pipeline and Shtokmanovskoye gas field development were listed as energy projects of common interest to Russia and the EU that were to be supported at the governmental level. This agreement was reached at the EU-Russia summit held under the EU-Russia Energy Dialogue programme, which aims to increase the reliability of Russian gas supplies to Europe[135].

---

[133]   The origins of the NEGP's idea date back to 1998 when Gazprom and Finnish Fortum agreed a 50-50 North Transgas JV to conduct a feasibility study.

[134]   *Western Gas Intelligence* (February 2002).

[135]   'EU-Russia energy dialogue making progress', *Gas Matters* (October 2002), p. 7.

Fig. 3.3: The North European Gas Pipeline

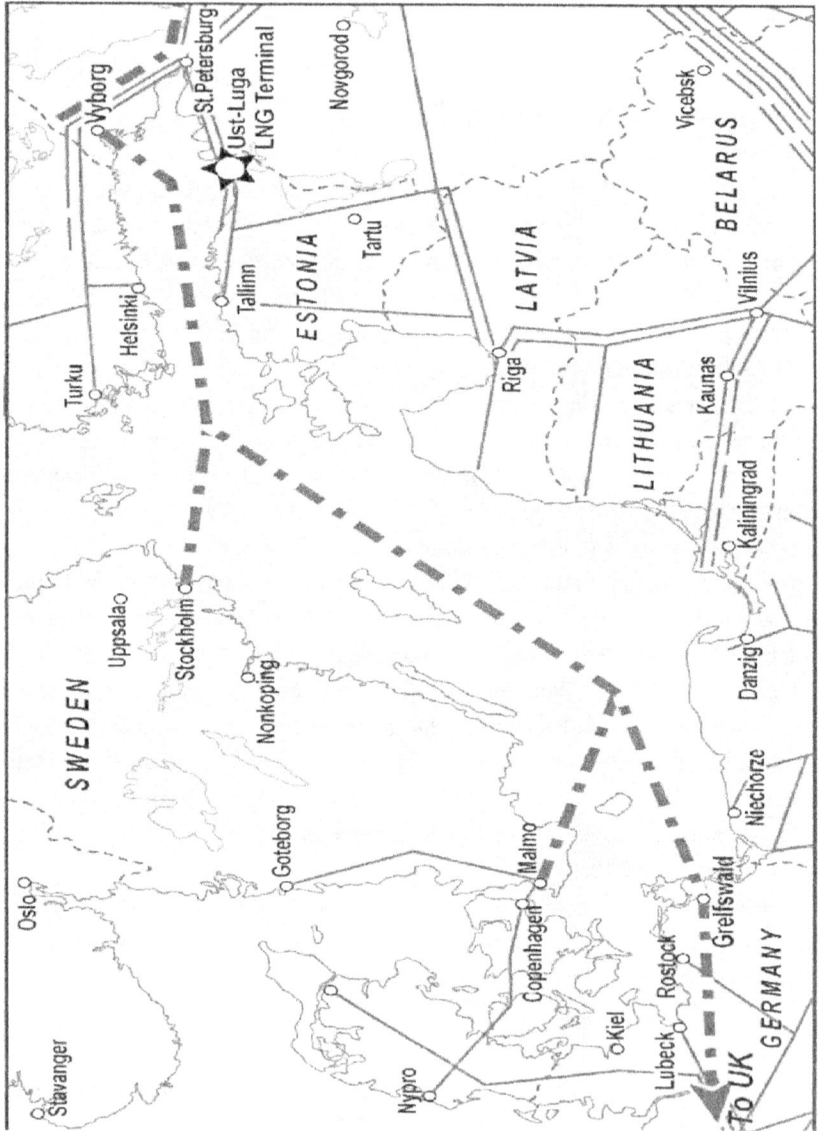

Source: Jonathan Stern, *The Future of Russian Gas and* Gazprom.

Some facts are suggestive that the decision about NEGP's construction has already been made. The Russian president has, for instance, repeatedly expressed his firm support for the project. However, some uncertainty about construction remains. Thus, for example, in February 2004 Gazprom stated that it aimed to make a final investment decision on the cost and pipeline routes in the fourth quarter of 2004. However, the timing of this declaration – February 2004, i.e., in the midst of Belarus-Gazprom conflict – is suggestive that it was partly motivated by a desire to 'bring Belarus into line' by threatening its leadership with the prospective loss of transit fees if the NEGP were built. Nonetheless, growing EU support and the enduring inability to conclude a long-term gas contract with Belarus in 2004 made Gazprom's board decide to approve an implementation of the NEGP project in November 2004[136].

The pipeline, projected to have seven compressor stations, would include a 568-km onshore section in Russia and a 1,089-km underwater section. Its capacity is envisaged to be around 30 bcm/year[137], although that figure cannot be anything but speculative until the final proposal is accepted (which has failed to materialise to date) and the firm European markets for gas to be supplied via the NEGP are established. The surveying had been envisaged to start in January 2003 with construction to begin in late 2004 or early 2005[138]. The works on the onshore part of the pipeline have been already begun, unlike the offshore works, which will only go ahead when a consortium is in place.

Agreements with potential partners for the NEGP's construction are crucial for Gazprom since the latter is unable to finance such a large project by itself. Even given that some European financial institutions, such as EBRD, agreed to finance a part of the project, without large companies' participation the funding available to Gazprom is inadequate. The EU's support of the NEGP in the

---

[136]   *Gas Matters* (December 2004). Gazprom started to develop Yuzhno-Russkoye field, which is to be one of source fields for the NEGP.

[137]   'Gazprom progresses on pipe schemes', *Western Gas Intelligence*, 13:47 (November 2002).

[138]   'EU-Russia summit swiftly followed by green light for North Transgas', *Gas Matters* (November 2002).

framework of the EU-Russia Energy Dialogue makes it less difficult for Gazprom to attract private companies to participate in the consortium. Indeed, several companies such as Wintershall, Rurhgas, and E.ON expressed an interest in the project[139]. Gazprom and E.ON announced an intention to cooperate on NEGP's construction[140] and agreed to create a joint venture to extract gas, construct a pipeline across the Baltic Sea, develop an infrastructure to supply gas to Europe and launch joint projects in the power sector of the European market. The Finnish government, whose interest is not surprising since Russia is the sole supplier of gas to Finland, also expressed its willingness to participate. The interest on the part of German companies may be explained by the fact that receiving 90% of Russian gas via Ukraine and 10% via Belarus, Germany wants to have an export route that would not depend on CIS transit countries.

The NEGP's existence would have important implications for all sides involved in the gas trade - Russia as a supplier, Europe as a consumer and Belarus as both a transit and consumer country.

Implications for Russia

Gazprom has good reasons to be very cautious about NEGP's construction having earlier undertaken a similarly massive investment in the Blue Stream pipeline passing under the Black Sea. As in the case of the NEGP, one of the major reasons behind Gazprom's decision to build the Blue Stream route was the desire to avoid unreliable Ukraine and Bulgaria and to secure uninterrupted gas supply to the Turkish market. However, the project proved to be a financial failure since Turkish gas demand was overoptimistic, and, in the event, Turkey does not contract enough gas to make Blue Stream a profitable investment. Indeed the pipeline is currently operating well below capacity. This

---

[139]   North Trans gas pipeline gathers momentum, *European Gas Markets*, 8:01:2 (January 2001).

[140]   www.MosNews.com. The announcement was made after the July meeting of the Russian President and German Chancellor.

book argues that bearing this experience in mind Gazprom should be quite cautious about any other Blue Stream-like investments.

There are serious considerations that make NEGP's construction look less attractive than it appears at first sight. NEGP is extremely costly to build and is far from being the cheapest route for Russian gas to Europe. According to Gazprom's estimations, the onshore pipeline costs $1 bln and the undersea one - $2.8 bln, adding up to $3.8 bln in total. An independent evaluation suggests a somewhat higher figure – $5-5.7 bln[141]. High costs mean that Gazprom's freedom of exports will be limited since it cannot afford to build the line without the participation of foreign companies, whose financing is conditioned upon equity/share in capacity. Another problem is that whereas the absence of unreliable transit countries on its way is the biggest NEGP's advantage, a serious disadvantage is that there are only very few intermediate gas markets along the route (such as Sweden, Denmark and Russian Kaliningrad) which are small and whose development prospects are limited[142]. As for the potential end-markets, there are also some questions - for example, it is far from certain that the UK market, which Gazprom believes will be an important buyer of Russian gas delivered via the NEGP, would not be more in favour of other gas suppliers. All this raises serious questions about the NEGP's cumulative net cash flows, profitability, and the pay-back period.

There is another danger, namely that Gazprom will not be able to ensure that its exports via the NEGP would not compete with the company's own gas flows through Belarus and Ukraine, thus causing a downward gas price pressure in Europe. Indeed, it is sometimes argued that the NEGP could disturb the current gas balance in Europe. Moreover, Gazprom itself admits that the European gas market is currently stable and additional gas volumes could destabilize it[143]. Thus the NEGP, instead of giving Russia an increase in earnings, may cause a reduction in income from exports. Whereas an additional

---

[141]   *Cedigaz*, 19 December, 2002
[142]   The economics of regional network transportation is only favourable if intermediate markets already exist or can be developed.
[143]   Gazprom uses the same argument when denying the 'independents' access to export markets.

gas export route from Russia westwards will become increasingly necessary in the future, in view of growing European gas demand, projected European demand for the next 5 years (by which time the NEGP if begun now should be completed) is not high enough to justify the NEGP's construction now. All these factors should make Gazprom think twice before embarking on the NEGP's construction.

That said, there are important advantages that the NEGP would bring to Gazprom. Not being an alternative for the existing transit routes, the offshore line nonetheless enables Gazprom to have a greater control and flexibility over a significant proportion of its day-to-day gas exports since it is projected that the line would transport around 20% of all Russian gas exports to Europe in the next 10 years. This, in turn, would allow for diversification and increased reliability of Gazprom gas exports, thus implying a growth of Russian budget revenues.

Implications for Belarus

The construction of the NEGP means a substantial economic loss for both Belarus and Ukraine. Even given that the NEGP's projected capacity does not enable it to handle all gas exports that currently flow via the Belarussian transit network, it would substantially weaken Belarus's bargaining power over transit fees and gas prices since a part of gas exports could be securely diverted via the NEGP. For the same reason, the offshore line would give Gazprom more flexibility in its usage of the Ukrainian gas network. Thus the NEGP is sometimes viewed as a means of bringing such transit countries as Belarus and Ukraine 'into line', because a real threat of gas diversion to the offshore pipeline represents additional leverage that Gazprom can exert over currently or potentially 'difficult' gas transit countries[144].

This book, however, argues that while weakening transit countries' bargaining power over gas prices and transit fees, the NEGP is not capable of enabling

---

[144]   'Газпром' хочет обойти стороной Беларусь и Украину', *Экономика и Бизнес*, Март 2005.

Gazpom to insist on its preferred financial terms. Indeed, a simple comparison of NEGP's total capacity, which is 30 bcm per year, and the volumes which Gazprom needs to export, which is about 150 bcm per year, suggests that the NEGP does not eliminate the existing interdependence between Russia and CIS transit countries though it does make it more asymmetrical in Russia's favour[145].

Therefore we argue that Gazprom, even if it does build the NEGP, will be forced to look for a compromise permanent solution to the conflictual issues with *either Belarus or Ukraine, or both - depending on the growth of European gas demand.* This is not only because the NEGP alone is not sufficiently large to handle Gazprom exports to Europe, but also because Gazprom still needs the CIS transit routes for diversification purposes. Therefore maintaining stable relations between Russia and Belarus, as well as between Russia and Ukraine, is necessary for securing the reliability of Russian gas supplies to Europe.

Implications for Europe

Since European gas demand is projected to grow, European gas companies are interested in the long-term addition of new gas evacuation routes from Russia. As long as these routes are directed towards Europe, however, there is no clear EU preference about any particular route[146]. Indeed, prior to the 2004 gas dispute between Belarus and Gazprom, the EU supported both the Yamal-Europe pipeline and the NEGP without indicating any preference towards either of them.

Being projected to connect European networks with the Russian gas system, the NEGP would considerably reduce the risks regarding the reliability of Russian gas exports to Europe. According to Gazprom, the NEGP is envisaged to

---

[145]   The current capacity of Yamal-Europe is 23-23 bcm (designed – 33), Northern Lights around 30 bcm, and the effective capacity of the Ukrainian network is 140 bcm.

[146]   'Russia's North Europe Gas pipeline moves', *Western Gas Intelligence,* 15:8 (February 2004).

transport around 20% of Russian gas exports to Europe within the next decade, which is equivalent to 26-30 bcm of gas annually. Thus the route increases the reliability of Russian gas supply, providing that Europe accepts a
high and prevalent dependence on Russian gas, as opposed to Norwegian,
Algerian and Dutch. However, given that European gas demand and Russian
gas exports to Europe increase roughly at the same speed, dependence on
Russian gas will not increase greatly[147].

The NEGP, being a co-operative project between Russian and the EU, can
only proceed fully when the problem of destination clauses[148] in existing long-
term contracts is resolved. Gazprom adopted a hard stance on the issue and
the negotiations regarding clauses' abandoning were slow and difficult. While
Gazprom agreed that all its *new* gas contracts would not include destination
clauses (these contracts are concluded predominantly with the 'old Europe'
countries)[149], it refuses to re-negotiate the destination clauses in the old contracts (signed in the 80s with the eastern European countries, then COME-
CON members)[150]. Therefore, for the EU-15 the destination clauses are not
much of an issue any more, and it is generally accepted that the country importing gas from Russia already sells gas further on, when domestic production allows this fact to be hidden. But for the accession countries destination
clauses will remain an issue for some time to come. This will not be easy to
negotiate since Gazprom, as we argued earlier, has a stronghold on gas supply in eastern Europe where many countries do not have an alternative supply
either because they cannot afford to diversify their gas imports or because it is
technically impossible[151]. However, since the destination clauses violate European competition law Gazprom will have to eliminate them sooner or later in
its east European contracts as well thus paving the way to fuller cooperation
with Europe on the NEGP's construction.

---

[147]   Jonathan Stern, *The Russian Natural Gas 'Bubble': Consequences for European
Gas Markets* (London: RIIA, Brookings), 1995, p. 72.

[148]   Clauses prohibiting to resell gas.

[149]   'Gazprom drops destination clauses', *European Gas Markets*, 10.10.1 (October
2003), pp. 1,7.

[150]   'EU accession-ten new countries, and a few new problems', *Gas Matters* (May
2004).

[151]   See p.4.1 for discussion.

4   How Significant Is Belarus's Gas Bargaining Power?

4.1   The Belarus-Gazprom Gas Dispute: Its History and Current State

The analysis carried out in the previous chapter demonstrates that the gas transit routes running through Belarus - Northern Lights and Yamal-Europe - constitute the important component of the gas pipeline network for Russian exports to Europe. The Belarussian routes have gained additional significance in view of recent changes in Ukraine that may well alter the rapprochement that occurred between Gazprom and Ukraine before the elections. This increased significance makes it important to explore the nature of the unresolved issues between Gazprom and Belarus in order to determine whether there are strong grounds for the Belarussian routes to regain their reliability.

This chapter analyses the history and evolution of Russian-Belarussian gas trade and transit during the last decade in the context of the 'unionist' tendencies assessed in detail in Chapter 1. We aim to explain why this previously reliable relationship deteriorated so far as to have culminated in Gazprom cutting the gas off in February 2004, thus undermining the reliability of Russian gas exports to Europe via Belarus. The book further analyses the prolonged post-crisis negotiations between Belarus and Gazprom in order to explain why the dispute is still not settled on a permanent basis.

Since the break-up of the USSR, Gazprom has been a major supplier of gas to Belarus, providing it with some 60-70% of its gas needs, with independent Russian gas companies covering the rest of the demand. Starting from July

2002, Belarus paid Gazprom just $17-18/mcm of gas[152], i.e., the same price as the Russian domestic price, whereas the average price for the CIS (including Ukraine) was about $50 and the border price for Europe – $100-124. This concession was due to the planned creation, envisaged in the April 2002 Intergovernmental Agreement, of a joint venture between Gazprom and Beltransgaz that would own and control the Belarussian transit and transmission network[153]. The plan was further confirmed by the Russian president during the January 2003 session of the Supreme Council of the Union of Russia and Belarus[154].

As it did in Ukraine and Poland, Gazprom sought a majority stake in the Belarussian network, insisting in summer 2003 on a 51% stake in Beltransgaz with Belarus, reportedly, declaring that it would not even agree to 25-30%. Nonetheless, the two sides later agreed on a 50/50 scheme[155]. This agreement confirmed the quantities of gas to be delivered, but gas prices and transit fees, as before, were to be stipulated by separate contracts signed between Gazprom and Beltransgaz. Apart from providing Gazprom with the means of control over the Belarussian network, the creation of Beltransgaz JV was seen as a way of settling Belarus's gas debt to Gazprom, which, according to the latter, stands, as of February 2004, at $120 mln, a considerable figure but still not a vast one compared with Ukraine's 1.5 bln in the 90s[156].

Gazprom sought to insure itself against political risk in Belarus by securing a clause in the agreement stating that any future Russian investment in

---

[152]   'Russia cuts transit flows through Belarus – Has Belarus become the new Ukraine?' *Gas Matters* (February 2004), pp. 9-10.

[153]   According to 'Russia and Belarus agree breakthrough gas deal', *Gas Briefing International* (June 2002), pp. 2-4, no deadline has been set for the implementation of the agreement. Gazprom, however, claims that the JV should have been formed by July, 2003.

[154]   'Russia and Belarus agree on joint transit venture on the way to economic union', *Gas Briefing International* (February 2003), p. 8.

[155]   The press-conference of Gazprom's deputy chairman, A. Ryazanov, and Gazexport's director general, A. Medvedev, Gazprom's website, 2.03.2004.

[156]   'EuralTransGas left out in the cold as Naftogaz Ukrainy and Gazprom settle their differences', *Gas Briefing International* (August 2004).

Beltransgaz would not be re-nationalised[157]. For its part, Belarus included a mutual clearance mechanism in the agreement, to write off $72 mln debt[158]. In exchange Russia would not pay the VAT and customs duties incurred during the construction of the Belarussian section of the Yamal-Europe pipeline. However, just a week after the agreement was signed, Gazprom raised objections to the clearing mechanism, and at the end of April 2002 Gazprom threatened to cut supplies to Belarus over its debt arrears for the 2002 deliveries.

Another complication arose about which assets were to be contributed to the venture. Belarus claimed that apart from the state-owned Northern Lights pipeline, the Belarussian section of the Yamal-Europe pipeline should be included whereas Gazprom was adamant that the latter should remain a separate entity because Gazprom had paid for its construction.[159] Also, Belarus refused to contribute the $0.5 bln to the JV agreed upon earlier.

But the most important disagreement concerned the valuation of Beltransgaz assets. Belarus declared that it valued Beltransgaz at $5-6 bln, and could not sell it at a book value of $500-600 - the price proposed by Gazprom. Later the Belarussian president lowered the price somewhat, suggesting $2.5 bln as a lower limit. Gazprom, however, refused the deal and said it would have taken 'more than half a century' to gain a return on such an investment.

Months after the agreement was signed Belarus still had not fulfilled its obligations, pretending, however, that the problem was technical rather than one of principle. In September 2002 Gazprom discovered that Beltransgaz was still on the national list of strategic companies that could not be privatised. This led Gazprom to threaten Belarus repeatedly with the suspension of its gas deliveries. This had the desired effect and, in November 2002, the Belarussian legislature, in notable contrast with the Ukrainian one, passed a bill allowing for the

---

157   'Russia and Belarus agree breakthrough gas deal', *Gas Briefing International* (June 2002), p. 4.

158   A penalty for late payments for gas supplied in 1997-1999.

159   'Belarus risks cuts in row over Russian joint pipeline venture'. *Gas Briefing International* (October 2002).

privatisation of Beltransgaz[160]. The government also agreed to settle its gas debt of $150 mln in 2003 but still argued that it should now be reduced as a result of a retroactive gas price calculation since the border price for Belarus decreased according to the April 2002 Agreement. Gazprom, however, wanted to cancel the discount since Belarus failed to fulfil its part of the agreement, i.e., the establishment of the JV. In 2003, witnessing no progress on the JV, Gazprom began to charge Belarus $30/mcm, as it did before the 2002 agreement was signed[161].

Price increase was a compelling argument, eventually making Belarus go as far as transforming Beltransgaz into a joint stock company in April 2003. The process, however, did not progress any further and later the government openly declared that it would not sell Beltransgaz because of the lowness of the proposed price. Gazprom repeated that it was not prepared to pay more than the book value and threatened Belarus with further price increases if the venture was not formed[162]. Importantly, neither the value of Beltransgaz nor even a mechanism for its valuation was fixed in the 2002 agreement when the sides agreed on the JV *in principle,* but not in detail. This circumstance en-abled both sides to raise all the objections listed above, which later proved impossible to overcome.

Thus when the 2003 gas contract between Beltransgaz and Gazprom expired, the latter demanded Belarus pay $50/mcm instead of $30. This step was backed by the Russian government, which cancelled its recommendations to the Federal Energy Commission[163] to sell gas to Belarus at Russian domestic prices[164]. Belarus refused to sign the 2004 contract on these terms. From 1

---

[160]   'Russia and Belarus (and Gazrpom and Itera) keep gas flowing – at a price', *Gas Briefing International* (November 2002), p. 2.

[161]   'Russia withdraws low gas price concession from Belarus', *Gas Matters* (September 2003), p. 3.

[162]   Gazprom claimed that its losses because of allowing Belarus to pay lower prices amounted towards $ 2 bln for the last four years.

[163]   Later renamed The Federal Tariffs Service.

[164]   'Russia now threatens a Belarus by-pass', *Gas Matters* (October 2003). The then Russian prime-minister, Kasyanov, had signed the order removing the concession from January 2004.

January 2004, in the absence of a contract, Gazprom stopped shipping gas to Belarus via Beltransgaz's Northern Lights pipeline, letting the independents – Itera, SIBUR and TransNafta - take up the business. They continued to supply Belarus but at a higher price than Gazprom – around $46.68/mcm – raising fears of possible increases to $50. These sales were happening under short-term contracts lasting one month or less. Gazprom, providing the independents with access to its pipeline network, waited for a successful conclusion to the negotiations with Belarus over the long-term gas contract foreseen in the 2002 Intergovernmental Agreement. On 18 February the last short-term contract expired, and the independents refused to conclude a new contract. This left Belarus practically without gas supply – except the gas that was still travelling through Belarus in transit to Europe.

Belarus, unlike Ukraine[165] or Germany, has almost no underground gas storage capacity that would enable it to withstand a stoppage of its external gas supply. The storage Belarus does possess is only sufficient for smoothing seasonal temperature fluctuations. Thus being 90% dependent upon gas imports, having most of its industry gas-fired and without gas storage, Belarus's position was absolutely critical[166].

However, the Belarussian government persisted in its refusal to sign a new contract with Gazprom to buy gas at $50/mcm. Instead, Beltransgaz, having notified Gazprom in advance but having not received its permission, undertook unauthorized gas offtakes from the Yamal-Europe pipeline, which is intended solely for gas shipments to Europe[167]. In response, despite Yamal's handling around 16% of its exports to Europe, Gazprom cut off gas flows via this route as well[168].

---

[165]   35 bcm.
[166]   TransNafta is not a gas production company and purchases gas on the Russian market.
[167]   In 2004 23-24 bcm of gas was transited to Europe via Yamal-Europe and 7-9 bcm via the Belarus-owned Northern Lights.
[168]   'Russian Gazprom stops flows to Belarus', *European Gas Markets*, 11:02:2 (February 2004), pp. 1, 6.

As our analysis suggests, contrary to the widespread notion that the February 2004 gas crisis was entirely unexpected, relations between Belarus and Gazprom had been deteriorating since 2002. Indeed, Gazprom started to question Belarus's reliability as early as 2000 - in convergence with the slowdown of political integration. Therefore the gas crisis seems to have been the logical culmination of mounting problems. However, its nature - a complete cut off of *all* gas flows via Belarus - was unexpected. This was the first time Gasprom has *ever* cut supply in this way. Even during the height of unsanctioned off-take of 10 bcm of gas in Ukraine in 2001, Gazprom did not go that far.

Having cut off gas flows via Belarus Gazprom inevitably stopped supply to those consumers in Germany, the Baltic states, Poland and Kaliningrad, which normally receive Russian gas via Belarus. By and large, the end-users in Europe did not notice a blip in supply because it was short-lived and supplies via Belarus were renewed on the following day, when the republic signed a new short-term contract with TransNafta, and also because Gazprom was successful in diverting gas flows via alternative routes (particularly via Ukraine). Germany was almost unaffected because 90% of its Russian gas imports normally flow via Ukraine[169]. Moreover, Germany has a substantial storage capacity – 18.5 bcm - that can be used to smooth even much larger supply disruptions[170]. Kaliningrad, the Baltics and Poland were most vulnerable to disruption. Gazprom maintained supplies to its Baltic consumers by having gas pumped to Lithuania through the previously unused Riga-Panevezys pipeline in Latvia – at 200 mcm per hour, plus a further 30 mcm per hour to Kaliningrad versus the combined 400 mcm per hour normally supplied to these consumers via Belarus[171]. Lithuania stated that these flows were sufficient for residential and major industrial needs, though it admitted to a brief interruption for some users. Poland reported more serious disruptions, which is not surprising considering that 70% of Polish gas imports are normally shipped via Belarussian routes, the Yamal and Northern Lights. The Polish gas company, POGC, had even threatened to demand compensation for

[169]   *Western Gas Intelligence*, 15:4 (January 2004).
[170]   'EU storage capacity', *Gas Matters* (November 2002).
[171]   'Gazprom image at risk', *Western Gas Intelligence*, 15:8 (February 2004), pp. 1-2.

the losses incurred because of a 5 million cubic meters shortfall in deliveries. However, the value of such a small shortfall can hardly be large, and Poland's step was politically rather than economically motivated.

The interruption pushed Poland to seek to diversify its gas imports. Poland held negotiations with Norwegian and Dutch gas companies regarding possible gas supplies, though no new contracts were signed, Russian gas being a much cheaper option[172]. Despite having failed to diversify gas imports, POGC, however, undertook alternative measures to increase the reliability of gas supply in the future by securing Gazprom's consent to build an underground storage capacity in Poland (1.2 bcm in the first 3 years, expected to reach 3 bcm) that would smooth the supply of gas through the Yamal-Europe pipeline[173]. Gazprom also agreed to install equipment that would allow gas flows to be reversed, i.e., to flow *from Europe,* to be delivered from Germany to Poland in case of further disruptions[174].

Overall, whereas the European Commission replied at the diplomatic level to the hiccough in supplies, no individual consumer-country, apart from Poland, made a formal claim to Gazprom[175]. However, although the February 2004 gas blip was very brief and caused no serious disruptions to European consumers, it nonetheless threatened Gazprom's long-standing reputation as a reliable supplier to Europe.

While explaining its decision to cut off supplies via Belarus, Gazprom, trying to save its reputation, stressed that only its 'unpredictable partner' – the Belarussian leadership – was to blame for gas shortfalls[176]. However this was not very confidence-inspiring for Gazprom's European consumers, since it did not give

---

[172]   *Ibid.,* p.2.
[173]   'Gazprom and POGC consider possibilities of Polish storage', *European Gas Markets,* 11:04:1 (April 2004).
[174]   'PGNiG plans to increase supply security following Belarus interruption', *Gas Matters* (April 2004), p. 17.
[175]   'POGC threatens to make claim over Gazprom-Belarus spat', *European Gas Markets (*31 March 2004).
[176]   See the report from the meeting of A.Miller and A.Ryazanov with Belarussian reporters, Gazprom's website, 17.05.2004.

them any indication of how the matter would be resolved and gave no guarantee that the shortfalls would not be repeated. Indeed, Europe, being a major consumer of Russian gas, cannot exert any meaningful pressure on Belarus to ensure the reliability of its transit routes, at least as long as Belarus remains under Russia's political and economic patronage. As for Belarus's reaction to the cuttings off, it was highly negative. The Belarussian president labelled them a 'terror act of a highest order' – marking the lowest ever ebb in relations between Russia and Belarus.

As a private company, Gazprom tried to distance itself from the state, stressing that even if a 'specific' relationship between Russia and Belarus existed, the relationship between Gazprom and Belarus as a consumer is nothing but business-like, i.e. no payments, no gas supply. However, while withdrawing price concession from Belarus in January 2004, Gazprom mentioned that its decision was sanctioned by the Russian government. This indicated changes in the 'specific' Russian-Belarussian relationship, marking a shift from their politicization to their economisation. Indeed, as early as September 2003, the Russian president stated that Russia and Belarus 'should switch to market relations in gas industry'[177], adding at the same time that the negotiations to form Beltransgaz JV must be continued. Encouraged, Gazprom further declared that it would only sign a long-term gas contract with Belarus after the JV was set up[178]. However, Gazprom's optimism remained unfulfilled and it had to sign the 2005 contract with Belarus despite the absence of the JV.

Less than a day after the cut off, Beltransgaz signed a new short-term gas contract - at $46.68/mcm - with TransNafta, to last until 3 March. In a similar way Belarus was supplied by the other independents until mid-2004. Meanwhile difficult and prolonged negotiations with Gazprom continued, and it was only in mid-June 2004 that the new contract was finally signed to ensure Gazprom's gas supplies to Belarus for the remainder of 2004. Russia agreed to

---

[177]   'Russia withdraws low gas price concession from Belarus', *Gas Matters* (September 2003).

[178]   'Belarus joins Russia's Open-Access Battle', *Western Gas Intelligence*, 15:2 (January 2004).

supply Belarus in 2004 with 18.5 bcm of gas, of which 8.3 bcm had already been delivered by the independents. Both Gazprom and the independents shipped gas to Belarus at $46.68 in 2004, and gas now flowed through Belarus at a higher transit rate - $0.75 per mcm /100 km, instead of $0.53 previously.

Mid-2004 marked some improvement in relations between Belarus and Gazprom, an important sign of which was the extension of the Russian government's $200 mln loan to Belarus to compensate it for gas price increases since 2004. As it was conditional on the signing of the 2005 gas contract, the loan thus came to effect in December 2004. Another positive sign was that Gazprom had recently increased gas transit through Belarus[179]. Furthermore, in trying to resolve the JV issue, the sides appointed an independent consultancy firm, London-based Deloitte&Touche, to evaluate Beltransgaz's assets.

The evaluation procedure was meant to be completed by the end of 2004, and the sides were meant to accept its results and finally establish the venture just before the 2004 contract expires[180]. This, however, did not happen – there is no news of the evaluation's progress, the JV is still not in place; and the talks between Belarus and Gazprom regarding the 2005 gas contract were no less difficult than those for the 2004 one. Gazprom tried to increase prices to $50/mcm whereas Belarus threatened to charge it higher transit fees - $1.04 per mcm/100 km. In the event, the sides agreed on the exactly the same terms as in 2004, i.e., gas prices remained at $46.68/mcm, and transit fees at 0.46 via Yamal-Europe line and 0.75 via Beltransgaz. However, the very fact that there *is* a contract makes Belarus better off economically because, while supply conditions remain the same, it now receives transit fees, unlike the first half of 2004 when Gazprom did not pay fees because of the absence of a contract[181].

---

[180]   See A. Miller and A. Ryazanov' interview, Gazprom's website, 17.05.2004.
[181]   The press-conference of Gazprom's deputy chairman, A. Ryazanov, and Gazexport's director general, A. Medvedev, Gazprom's website, 2.03.2004.

This book argues that the gas conflict between Belarus and Gazprom is still not resolved on a permanent basis since the signed contract is of a short-term – 1 year - nature. In the absence of a long-term contract the dispute could easily bounce back in 2006, as already happened in 2004. Indeed Gazprom has already declared that it intends to charge Belarus higher gas prices in 2006. Belarussian officialdom, for its part, voiced its intention to seek an increase in gas transit fees. The situation is reminiscent of 2004 when talks ended in a stalemate and Gazprom decided to cut off gas flows via Belarus. The book, however, argues that the current situation is different in one important respect, namely that Gazprom cannot risk further damage to its reputation by cutting off its gas exports to Europe for a second time. Therefore Gazprom and Belarus have good grounds to be interested in a mutually acceptable commercial solution rather than all-out conflict.

4.2   Factors That Push Belarus and Gazprom to Re-negotiate

The book further assesses the 'cards' that Belarus, as a gas consumer and transit country, and Gazprom, as its major and only - from 2005 - gas supplier, hold in order to negotiate the gas conflict resolution.

Gazprom has important leverage in its bid for Beltransgaz because Belarus, having almost no indigenous gas production[182], needs gas and can only realistically buy it from Russia. Even if Belarus could buy gas from Turkmenistan, this would still need to travel through Gazprom's pipelines. Importantly, the same applies to independent Russian gas companies, the issue which we will look at in more detail later in this chapter.

As our analysis in Chapter 2 suggests, Belarus with its unreformed economy cannot afford to buy gas at increased prices without substantially cutting its numerous subsidies and thus triggering political instability. This is especially true because Gazprom now insists on cash payments, as opposed to its earlier acceptance of barter. Thus only if Belarus undertakes a genuine restructuring of its industrial complex, enabling gas conservation and efficiency measures to be implemented, an increase in national competitiveness and a decrease in gas demand will improve the country's ability to pay higher prices for gas imports.

Nonetheless even in the current situation, Belarus still has some important 'cards' to play because Russia needs the Belarussian routes for its gas exports to Europe. Reportedly, opinions are divided at Gazprom over whether it needs the Northern Lights pipeline, which currently transits Russian gas not passed through the Yamal pipeline[183]. The argument, which is made against the acquisition, is that the Belarussian system, while in a better condition than the Ukrainian one, is seriously ageing in parts and therefore needs considerable investment. This book argues that the ageing Belarussian infrastructure

---

182   0.3 bcm.
183   'Russia and Belarus agree breakthrough gas deal', *Gas Briefing International* (June 2002), p. 4.

is, in fact, an argument in support of its acquisition because otherwise the line will not be usable for transit since it will deteriorate completely as a result of Belarus's inability to refurbish it. Furthermore, the refurbishment should be done promptly because if the line is allowed to deteriorate its refurbishment cost will increase exponentially. Gazprom, for its part, having acquired the network, would be willing to invest in its maintenance, thus contributing towards its higher technical reliability.

However, the main reason Gazprom would want to form the Beltransgaz JV is its desire to run, jointly with Belarus, the Belarussian section of the Yamal-Europe pipeline. Even if it would not be able to stop offtakes, it would make it easier for Gazprom to track them and force Belarus to pay later, as was the case with Ukraine in 1998-2000, thus increasing the reliability of gas transit.

The major reason Gazprom's European gas consumers did not have a major supply disruption in February 2004 is that the disruption was brief. Had it been longer, Gazprom might not have been able to maintain an uninterrupted supply to Europe. Certainly, Gazprom cannot sustain uninterrupted supplies to Europe *on an annual basis,* as opposed to short-tern diversions, using only the Ukrainian transit network. The latter cannot be utilised to its full capacity due to its poor technical condition. Furthermore, in 2004, Gazprom was able to use the Ukrainian network fairly easily to divert gas exports from 'rebellious' Belarus due to a notable improvement in relations between Gazprom and Ukraine in 2002-2004, an improvement that occurred against a background of deteriorating relations with Belarus.

Since it is unclear whether this rapprochement will continue in post-December 2004 Ukraine, Gazprom cannot be sure about its ability to use the Ukrainain network for gas diversions as easily as it did in 2004. Therefore Gazprom will be averse to cutting off supplies via Belarus again since it might result in more serious disruptions, thus damaging Gazprom's reputation. Even if Gazprom retains its ability to use the Ukrainian routes freely, a diversion of *all* gas flows from Belarus through the Ukrainian network is more of a last resort strategy than a solution to the Belarussian transit problem.

Only if the Ukrainian lines are eventually refurbished could they enable Gazprom to divert gas flows steadily away from Belarus. But as we argued at length in Chapter 3, the refurbishment of the Ukrainian transit network would not be a trivial task. An international consortium created by Gazprom and Ukraine in late 2004 brought some hope that the refurbishment would be accomplished, but after Ukrainian elections the consortium's fate is unclear. Since the EU is interested not only in increasing the reliability of the Ukrainian routes to decrease its dependence on Belarus but also in securing democratic achievements in Ukraine, it would be willing to lend its support to the consortium *with the participation of a third party*, thus providing a counterweight to Russian influence. However, more than support is needed, i.e., money which European institutions will not provide and which can only come from international energy companies. Their willingness to invest in Ukraine is hampered by uncertainty regarding property rights[184].

If the consortium is eventually established, and the necessary refurbishment of the Ukrainian network is carried out, it would admittedly decrease the significance of the Belarussian routes for Russian exports to Europe. However, transiting gas to Europe via Belarus is by far the shortest and most profitable route for Gazprom. Furthermore, Gazprom cannot just stop using the Yamal-Europe pipeline passing through Belarus and give up the huge investment already sunk into its first string. Therefore, we argue that Gazprom would still be willing to establish the JV to operate the Belarussian routes.

Belarus, for its part, being dependent on Russia for far more than just gas and, indeed, facing inevitable gas price increases which its unreformed economy is unable to withstand, is also interested in renewing negotiations over the JV in order to secure prices at least as low as domestic Russian ones[185]. This book argues that both sides, Russia and Belarus, are likely to return to the negotiating table and be more willing to re-negotiate the price of Beltrans-

---

[184]   Also, participation of foreign companies could be a crucial factor in providing confidence to the international lending community. See Jonathan Stern *The Russian Natural Gas 'Bubble'*, p. 72.

[185]   These prices being much lower than prices for the CIS and, indeed, for Europe, nonetheless, are much higher than they were back in 2003.

gaz. However, the timing of such negotiations is extremely important since if the NEGP – the offshore pipeline bypassing Belarus and Ukraine – is built in the meantime, it would reduce Gazprom's dependence on Belarussian routes and therefore worsen Belarus's negotiating position vis-à-vis Gazprom. Furthermore the NEGP would provide Gazprom with important leverage not only over Belarus but also over Ukraine, thus enabling Gazprom to secure its place in the international consortium to operate the Ukrainian routes. This, in turn, further weakens Belarus's position giving Gazprom more freedom to operate the Ukrainian routes. Therefore Belarus is in a better position to negotiate with Gazprom over the Beltransgaz JV now, before the line is built. Since it would take several years to construct the NEGP, which will not be completed before 2010 even if begun in 2005, Belarus has some time for negotiations.

In spite of the tremendous cost and possible creation of other by-product problems, discussed at length earlier, the NEGP's construction looks all the more likely as the lack of clarity and progress in the Beltransgaz JV establishment made EU institutions support the NEGP as a means of reducing European transit dependence. However, the offshore line would not give Gazprom enough capacity really to abandon the Belarus and Ukrainian routes. Also, because Gazprom needs them for diversification purposes, it is still bound to look for long-term solutions to conflictual gas issues with CIS transit countries.

4.3   Belarus and the Role of the Russian Independent Gas Companies

It is only since 2004 that Gazprom has supplied Belarus with 100% of its gas needs. Prior to that and since the USSR break-up, Belarus was supplied - apart from Gazprom, - by a number of independent Russian gas producers and traders (the independents) such as Itera, Sibur and TransNafta, which at different times covered around 1/3 of the republic's total gas needs. The enduring presence of the independents on the Belarussian gas market and the important role that they played supplying Belarus during the 2004 gas conflict when Gazprom stopped gas flows to Belarus, makes it important to explore whether having the independents as additional gas suppliers improves the position of Belarus as a gas consumer.

The common feature of Russian independents is that in order to transport their gas to the consumer they need to apply for access - the third party access, TPA - to the domestic pipeline of which Gazprom is the sole owner. Despite TPA being guaranteed by law and enforced by regulatory agencies, Gazprom still has a substantial power to influence the TPA and deny it on grounds that are difficult to prove, such as, for example, the lack of capacity in the pipeline, thus making the TPA partially discretionary. Importantly, the TPA rights are much more discretionary to CIS markets than to the domestic gas market.

In 1997 the independent Russian gas producer[186] and trader, Itera, started to supply the Belarussian market and became its largest independent gas supplier, its production being large enough to cover around 30-40% of Belarus's total gas needs[187] (see Table 4.1).

---

[186]   Unlike Sibur and TransNafta Itera has its own gas production.
[187]   'Gas trader Itera clocks up further sales in the fSU', *European Gas Markets*, 7:04:2 (April 2000), p. 7.

Table 4.1: Dynamics of Exports of Russian Gas by Gazprom and Itera to Belarus, bcm

	1998	1999	2000	2001	2002	2003	2004	2005
Gazprom	14.7	12.2	10.8	16.2	10.2	10.2[188]	10.2	19.1[189]
Itera	1.1	4.3	5.8	5.1	5.9	6.3	6.4	-

Source: Gazprom, *EGM, WGI.*

Itera's rise happened in the 90s under the previous Gazprom management, which sold Itera separate gas fields and parts of gas fields, for whose development Gazprom did not have enough finance at the time. As the new management was appointed to Gazprom in 2001[190], numerous allegations surfaced that the former Gazprom executives personally benefited from insider dealings with Itera[191]. The latter came under serious attack, and Gazprom continues to distance itself from Itera to prove to its shareholders that it deals effectively with the legacy of corruption[192]. S. O'Sullivan suggests that Gazprom prepared even to dismantle Itera completely rather than allow the previous insider dealings between itself and the latter to become public.

Indeed with changes in the Gazprom's policy towards Itera, the latter's production began to shrink, as being denied a previously reliable TPA it sold its several gas fields. Thus starting from 2002 Gazprom, in violation of the agreement achieved in 1997, continued to deny Itera access to Zapoliarnoye-Urengoi pipeline without which Itera cannot move its gas from the Beregovoye field[193]. Unable to transport gas, Itera was forced to sell its several producing

---

188   See footnote 84 for clarification.

189   Volumes contracted.

190   A.Miller, deputy energy minister and a member of pro-reform part of Putin's team, replaced R.Viakhirev in 2000.

191   The unequivocal evidence of that is still lacking since Gazprom shows lack of cooperation with PWC, which has been auditing Gazprom for 7 years.

193   'Gazprom blocks Beregovoye pipeline access to Itera', *European Gas Markets,* 10:05:2 (May 2003), p.8. See also 'Gazprom continues to deny Itera access to Beregovoye', *European Gas Markets,* 11:02:1(February 2004), p. 11.

fields back to Gazprom and also to Novatek. For example, in 2002 Itera sold 32% of its production affiliate Purgas to Gazprom[194]. As a result, during just one year Itera's production declined from 23.3 bcm in 2002 to the 15 bcm level in 2003. In 2004 Itera offered to sell its 49% share in the production company Achimneftegas (based on 353 bcm Urengoi field)[195] to Gazprom and also wanted to sell its production assets in Beregovoye. By 2004 Itera lost most of its fields, being left with only the unconnected Beregovoye field and a share in the Gazprom-controlled Purgaz.

Forced by Gazprom, Itera's decline in production weakened its position on the Belarussian market since it now had less spare gas to ship to Belarus[196]. By mid-2004 Itera completely departed from Belarus, and Gazprom took over, having added to its exports the remaining 38% of the market previously held by Itera, thus becoming the sole gas exporter to Belarus.

Pushing the independents from the CIS markets is the relatively new policy of Gazprom, and it is not limited just to Belarus. Indeed, Gazprom has been consistently *crowding out the independents from all the CIS markets*, because it itself wanted to benefit from their improved payment discipline. Indeed, Belarus, for example, was from 2003 paying Itera predominantly in cash, as opposed to its earlier barter payments. Until very recently, 2003, when the majority of gas trade with Belarus was concluded by barter[197], Gazprom itself put the Belarussian gas sales in Itera's way because they were loss-making for Gazprom.

The reason why the Belarussian gas market was desirable for Itera, was that not being limited by the Federal Energy Commission's price regulations, Itera charged Belarus $30-40 in 2002 and around $46.68/mcm in 2004 thus, actu-

---

194   'Itera offers shares in production company to Gazprom', *European Gas Markets* (14 May, 2004), p. 12. Gazprom also considered recovering form Itera the 32% stake in the Gubkinskoye field in 2001.

195   'Itera faces downsizing by Gazprom – again', *Western Gas Intelligence*, 15:20 (May 2004),p. 4.

196   'Gazprom rekindles its co-operation with Itera', *European Gas Matters* (June 2003).

197   In 1997 only 8% of Belarussian payments to Gazprom was in currency

ally, making a profit in Belarus. Whereas Gazprom shipped gas to Belarus at a loss at prices regulated by the FEC, i.e., $17-20 in 2002 and $30 in 2004, - due to the Customs Union and the Intergovernmental Agreement arrangements. Also, Itera, unlike Gazprom, was a highly efficient barter trader[198] by virtue of its barter expertise and also because of tax advantages that Gazprom, unlike the independents, did not have. Therefore, Gazprom was quite happy that Itera supplied Belarus with around 1/3 of its gas needs, thus saving Gazprom from additional losses that the latter would have incurred had it covered all of Belarus's gas demand instead of just 2/3 of it.

Thus being interested in keeping Itera in Belarus in 2003, Gazprom helped Itera to close up its balance to prevent its 'inopportune' departure – from Gazprom's perspective - from the Belarussian market[199]. The situation was that Itera, having lost many of its gas fields by 2003, was unable to supply both Belarus and Sverdlovsk region in Russia. By only cutting gas supplies to Belarus Itera could free the necessary 6.4 bcm of gas that it needed to cover all of the Sverdlovsk region's needs. However, Gazprom interfered and transferred to Itera the necessary volumes, thus enabling the company to honour its obligations in Belarus as well.

The situation changed in 2004 when the Russian government allowed Gazprom to charge Belarus higher prices for gas because of the Belarussian government's failure to establish the Beltransgaz JV. We argue that the newly acquired right to charge prices that would cover cost and profit was the major factor that made Gazprom revise its export strategy towards Belarus and oust Itera from the Belarussian market.

As we argued in Chapter 2, the price of $46.68/mcm of gas being already too heavy a burden for the Belarussian economy is bound to increase further be-

---

[198]   Itera shipped gas directly to the several Belarussian companies such as Naftan and Belaz car factory and invested in various Belarussian enterprises including meat processing and livestock feed manufacturing.

[199]   'Russia and Belarus (and Gazprom and Itera) keep gas flowing – at a price', *Gas Matters* (November 2002), p. 2.

cause Russia itself continues to increase its domestic gas prices, aiming at doubling them by 2010. Whereas allegedly Belarus's aim in its negotiations with Gazprom is to secure gas supplies at Russian domestic prices, Belarus, in effect, *is already buying Russian gas at domestic prices*, $46.68 being roughly a current gas price for the Russian 9th zone's consumers plus transportation cost. We argue that what Belarus *really* wants to achieve is to buy gas not at the *Russian domestic price* but at the *low price* - as it was back in 2002, when it was $17, or at least in 2000, when it was $30. However, this is unrealistic because Russian prices are already higher than $17-30. Nonetheless, the Belarussian government still believes that if it finally agrees to establish the Beltransgaz JV Belarus would receive Russian gas cheaply yet again. The only guarantee that the JV would provide, this book argues, would be for Belarus to receive gas at domestic Russian prices, i.e., 46.68 plus a 20% annual increase (the rate at which Russian domestic prices themselves are envisaged to grow), which, while being much lower than the border price for Europe, is not as low as Belarus would expect and, indeed, already too high for the unreformed Belarus to afford. Furthermore, if the JV is established and a long-term gas contract between Belarus and Gazprom is signed, Belarus's ability to increase transit fees in its framework would be limited because the latter will become a commercial, and not a political matter. Then the economic considerations, and not a whim of political leadership, will be a major factor influencing changes in prices and fees.

Whereas prices for Belarus cannot increase more slowly than domestic Russian prices, they may increase more rapidly. This book argues that Belarus cannot prevent gas prices from rapidly increasing unless either: A) the Beltransgaz JV is established; or B) the TPA to ship gas to the CIS becomes non-discretionary thus providing for the genuine price competition between Gazprom and the independents and, consequently, for less rapid price increases.

Ironically, being the only supplier of gas to Belarus gives Gazprom less flexibility. Indeed, having Itera and also TransNafta and Sibur as the suppliers 'of last resort' back in 2004 was convenient for Gazprom because it allowed it more time for negotiations with 'difficult' Belarus during the gas crisis and to avoid further unauthorised gas offtakes from the Yamal before the contract was finally signed in mid-2004. It also enabled Gazprom to take a firm stance on Belarus and to demonstrate that further gas price increases are inevitable because as early as 2003 Itera charged Belarus with $46.68/mcm (cost + profit) as opposed to Gazprom's $30/mcm (subsidised prices below cost).

We argue that, having pushed Itera away from the Belarussian market, Gazprom would only want to remain the sole supplier if Belarus, in fact, is able to pay the $46.68/mcm required by Gazprom plus the 20% annual increase. Otherwise Gazprom would be willing to hand over a part of its Belarussian exports to the independents as was the case prior to 2004 with some other independent company taking Itera's place. Such a desire would be strengthened further by the fact that, having had none of its (non-European) markets paying high enough prices to make investment in the development of new fields possible, Gazprom wants to free additional gas volumes from badly-paying Belarus, to deliver them further to the European market, or to the domestic Russian market which for the first time became profitable in 2004 [200].

However, since the financial position of independents others than Itera strengthened significantly during recent years despite the threatening rhetoric[201], and their TPA to ship gas *domestically* in Russia, as opposed to the CIS markets, has increased hugely[202], they might also prefer to serve domestic consumers rather than take on the unprofitable Belarussian market - even if the TPA to serve the CIS will become non-discretionary. Thus a decision about which market to supply first will be taken on the basis of the *economic performance of the potential gas consumer*. Therefore, if the Belarussian

---

[200]   'Gazprom to make domestic profit in 2004', *European Gas Markets,* 11:03:2 (March 2004), pp. 1, 6.

[201]   In October 2003 Gazprom threatened all the independents with a halt to pipe access and in September 2004 Gazprom demanded a 20% increase in TPA tariff.

[202]   Putin pointed out the need for independents to obtain equal access to gas pipelines but stresses that it 'does not mean that the independents will get free access to international markets'. 'Putin on Gazprom', *Western Gas Intelligence,* 15:11 (March 2004), p. 5.

economy remains unreformed then its attractiveness for the independents will decline in comparison to that of the domestic Russian market.

Furthermore, if the liberalisation of the Russian gas market progresses as far as to introduce the non-discretionary TPA for independents' gas shipments *outside the CIS*, which is currently *de facto* non-existent and is Gazprom's sinecure, then the Belarussian market is in an even bigger danger of being avoided by the independents. There was indeed some belief that the Russian government would introduce European export quotas for the independents once increased export pipeline capacity had been added through Yamal-2. But, as we argued earlier, Yamal-2's prospects look slim because Gazprom no longer perceives the Yamal route through Belarus as a reliable export corridor. Thus with Gazprom's current pipeline capacity westward being limited, Russia, despite the continuous EU pressure, is unlikely to end Gazprom's monopoly on European exports. Consequently, the independents are forced to limit themselves to the domestic Russian (and possibly CIS), market thus making the position of Belarus less difficult since it will not have to compete with the European markets but only with the domestic Russian market for the independents' gas.

Conclusion

This book contends that the relationship between Russia and Belarus strongly affects the reliability of Russian gas exports to Europe, since a significant volume of Russian gas passes through Belarus, whose role as a gas transit country has become increasingly important as a result of construction of the Yamal-Europe gas pipeline system. Since the Yamal-Europe system interlocks Russia, European consumers and Belarus in the gas trade and transit, it increases the vulnerability of Russia (as a gas supplier) and of a set of European countries (as gas consumers) to the possible political hostility from Belarus as a gas transit country. As demonstrated by the ongoing disputes between Belarus and Gazprom which resulted in a complete cutting off of gas flows via Belarus for a short period in 2004, the Russian gas exports to Europe to some extent become hostages to Russian-Belarussian relations. Furthermore, since Belarus is not only a gas transit country but also a gas consumer itself, and one that has enduring payment difficulties, it, in turn, has become hostage to its relations with Russia.

The initial impetus to Russia's growing decisiveness into dealing with political and economic issues separately was given by the 1998 financial crisis, which made it clear that the Russian economy could no longer maintain Russia's aspirations to superpower status. Following Putin's accession to power Moscow began to attach even more importance to the economic aspects of integration as opposed to the geopolitical ones. It is noteworthy that despite the growing divergence of Russia and Belarus' political regimes – towards more authoritarian positions – Russia, while continuing to lend its support to the Belarussian regime, is unwilling to support the ineffective Belarussian economic system. Indeed, by increasing gas prices for Belarus, Moscow forces the Belarussian leadership to introduce economic reforms that can enable Belarus to pay for Russian gas imports.

The book argues that the relations between Belarus and Russia, which seemed genuinely amicable for nearly a decade after the break-up of the USSR, were in fact much more complex. As long as there was a serious mutual intention of political integration between Belarus and Russia, the gas issues were amicably settled within the political dimension without paying much attention to the commercial side of the deals concluded. Indeed, a political union was initially seen in Russia as the reliable means of eventually gaining full control over the Belarussian transit routes, thus solving the problem of gas transit. However, with Putin's accession to the presidency the political integration process cooled, and now there is scarcely any common ground left between Russia and Belarus on political unification. While Belarus is reluctant to be politically subordinate to Russia, Russia is unwilling to have Belarus as an equal political partner in the union.

The idea of political union was seriously downgraded in both countries as a result of the growing 'economisation' of Russian foreign policy. This made the existing cleavages and unresolved economic problems in the gas sphere more visible. By 2000 Gazprom, having witnessed the growing divergence of the two countries' attitudes towards political integration, realised that Belarus's record of non-payment and gas debt accumulation would eventually undermine the reliability of Belarussian transit routes as had been the case in Ukraine in the 90s.

Therefore Gazprom, anxious to keep and expand its share of the very profitable European gas market, tried to prevent a repetition of the Ukrainian scenario by attempting to establish the Beltransgaz joint venture to own and operate jointly the Belarus transit network, ensuring an uninterrupted transit of gas to Europe. Belarus, having initially agreed, soon realised that selling its gas network in the absence of a political union would mean losing the only leverage available to the republic to influence the integration process and therefore reneged on the agreement. Since the establishment of such an agreement, together with the Customs Union agreement, was the main condition of why Belarus had since received gas at domestic Russian prices, reneging on it meant that Gazprom, backed by the Russian government, was able to abolish price discounts and require prices increase – from $17-30 in 2002-2003 to $50/mcm in 2004.

Attempting to keep gas prices low while refusing to privatise Beltransgaz, the Belarussian government tried to shift the gas issues from the economic to the military dimension, thus, for example, threatening Russia with charging it for the two military sites located in Belarus and for PVO service in violation of previous agreements. However, Belarus's attempts to use military issues as 'bargaining chips' in the gas conflict are bound to fail because military cooperation is beneficial to both countries, not only Russia, and because by acting in such a way Belarus might harm its own interests.

Because of a lack of power to influence Russia's stance by bargaining in political and military dimensions, Belarus is forced to resolve gas issues in the economic dimension. But in order to be able to act meaningfully as an independent economic actor, while refusing to establish the JV as a condition of continuing Russian domestic prices and a way of debt settlement, Belarus must be able to pay the demanded gas prices. This book contends that the unreformed Belarussian economy, in the absence of external and insufficient domestic financial sources, is too weak to withstand the pressure of increased gas prices, and that apart from the Russian loans (which were provided because Gazprom was trying to settle the conflict peacefully without resorting to a yet another gas war as in 2004) Belarus does not have enough other sources to pay for more expensive gas imports. Thus Belarus's independent stance vis-à-vis Gazprom, i.e., accepting higher gas prices rather than selling off its transit network does not have sufficient economic grounds.

In fact, even current gas prices - $46.68/mcm – are already too heavy a burden for the largely unreformed Belarussian economy. However, $46.68/mcm is the very minimum that Belarus should expect to pay Gazprom because it is, in effect, *already equal to Russian domestic gas price plus the transportation cost*, and Russia will not sell gas to Belarus at prices that are lower than for its domestic consumers. Moreover, gas prices for Belarus are bound to increase further because Russia itself increases its domestic gas prices at 20% per year according to Russia's own energy strategy and in line with WTO requirements. Thus even the JV establishment would not secure gas prices for Belarus at the low level of 2002-2003. The JV would only secure gas deliveries for Belarus at Russian domestic prices, which are still much lower than the European border price but nonetheless not as cheap as Belarus was accustomed to prior to 2004. Furthermore, if the JV is not established there is no guarantee for Belarus that the gas prices will not increase at an even higher rate than in Russia, thus making it even more difficult for Belarus to pay for Russian gas imports.

Another possibility for Belarus to secure less rapid prices increases in the absence of the JV could be achieved as a result of price competition between Gazprom and the independent Russian gas companies. However, such competition would only be genuine if the TPA rights to CIS markets eventually became non-discretionary. So far the independents are confined to the Russian domestic market, and the prospects of their full-fledged presence on the CIS markets, not to mention the European market, are questionable. Therefore as long as TPA to CIS markets remains discretionary, with Gazprom deciding which companies have rights of access, the only possibility for Belarus to prevent gas prices from increasing more rapidly than Russian domestic prices is to form the JV with Gazprom.

This book contends that Belarus's bargaining position vis-à-vis Gazprom over gas prices, transit fees and conditions of JV establishment is seriously influenced by changes in the reliability of the Ukrainian network. As soon as Ukraine's gas payment discipline improved and its relations with Gazprom had entered a friendlier phase following the 2004 settlement of Ukrainian 1997-2000 gas debt, the Ukrainian network became more reliable for Gazprom's European exports, thus making the Belarussian routes somewhat less important. At the same time, however, in view of political uncertainty in the aftermath of the December 2004 elections in Ukraine, Gazprom cannot be sure whether it will be able to use the Ukrainian network as easily as it did in 2004 to divert gas exports from 'rebellious' Belarus and avoid more serious disruptions to the supply to its European consumers. In any case Gazprom cannot sustain uninterrupted supplies to Europe *on the annual basis* without using the Yamal route since the Ukrainian network is in a poor technical condition, and cannot be utilised at its full capacity. The refurbishment of the Ukrainian network would enable Gazprom to divert gas flows away from Belarus on an annual basis. However, due to political uncertainty the future of the Gazprom-Naftogaz Ukrainy international consortium, which was established with the main purpose of maintaining the ageing Ukrainian network, is unclear. The consortium may still go ahead with some non-Russian foreign energy company participation, which would provide for a more politically acceptable framework of energy interdependence between Ukraine and Russia. However as long as property rights in Ukraine remain uncertain, the willingness of foreign companies to participate is limited despite the support that the EU lends to the consortium as a counterweight to Russian influence in Ukraine.

Moreover, even if Ukraine's network is eventually refurbished, thus decreasing the significance of the Belarussian routes, Gazprom would still be willing to use the routes passing via Belarus, if not as a strategic export corridor then as part of Gazprom's diversification strategy. These routes remain the shortest and thus the cheapest path for Russian gas exports to Europe. This book argues that the same logic applies to the possible construction of the offshore NEGP bypassing both Belarus and Ukraine. The NEGP, reducing Gazprom's dependence on CIS transit routes by adding 30 bcm to westward export capacity, would make interdependence between Gazprom and Belarus more asymmetrical in Russia's favour but, nonetheless, would not eliminate it completely; Gazprom would still need Belarus's routes for diversification purposes.

Therefore, since Gazprom needs to look for long-term solutions to gas conflicts with Belarus, and Belarus needs to secure gas supplies at prices not higher that domestic Russian prices, this book contends that both Russia and Belarus are likely to return to negotiations over the Beltransgaz JV establishment. Joint operation of the Belarussian routes, via the JV, would make it easier for Gazprom to monitor unauthorised gas offtakes (and force Belarus to pay later), thus increasing the reliability of gas supply via Belarus.

However, the major reason for the unreliability of gas transit via Belarus – the national economy's weakness resulting in debt accumulation and non-payment - would not simply be eliminated with the formation of the JV. Unless Belarus undertakes reform of its inefficient energy-intensive industry and introduces gas saving measures enabling it radically to reduce its gas consumption and improve its ability to pay, the republic will be tempted to resort to unauthorised gas offtakes and non-payments - which the JV could not prevent - thus undermining the reliability of Russian gas exports to Europe. Therefore the economic reforms in Belarus are the key factor to making its gas transit routes more reliable.

However, there is an important trade-off between Belarus's ability and willingness to embark on reforms. The Belarussian government is unwilling to introduce reforms because it fears that they would lead to the deterioration of living standards and thus political destabilisation. However, this book argues that since political stability rested on the economic welfare earlier provided by Russian energy subsidies, it was bound to weaken anyway as soon as Russia increased gas prices and started to insist on currency payments. Therefore, the Belarussian government in fact, no longer has a choice about whether or not to reform, because in the absence of the Russian economic patronage not reforming is as dangerous for the regime's stability as reforming.

However, even if there were a desire to reform on the part of the Belarussian leadership, its abilities to implement market reforms are limited. Indeed, even the Energy Saving Programme adopted by the Belarussian government proved to be difficult to implement in a command economy where price signals are substituted by less effective administrative controls. Furthermore, the specific nature of the Belarussian leadership hampers Belarus's chances of joining the WTO, a move which could provide Belarus with a powerful stimulus for further reforms. Indeed, since both the EU and the US perceive market economy and democracy as a package, it is impossible for Belarus to join the WTO as long as its regime is perceived as undemocratic.

Thus this book concludes that widespread perception of the Belarussian political regime as undemocratic weakens Belarus's chances to improve its economic performance as a necessary condition of making this country's transit routes more reliable for Russian gas exports to Europe.

Bibliography

1. Roy Allison, 'Russia and the New States of Eurasia' in Archie Brown, ed., *Contemporary Russian Politics: A Reader* (Oxford: Oxford University Press, 2001). pp. 443-452.

2. Ewan Anderson, *An Atlas of World Political Flashpoints: A Sourcebook of Geopolitical Crisis* (London: Pinter Reference, 1993).

3. Maarten J. Arentsen and Rolf W. Kunneke, *National Reforms in European Gas* (Elsevier, 2003).

4. Raymond Aron, *Peace and War (*New York: Doubleday, 1996), p. 191.

5. Margarita Balmaceda, 'Myth And Reality In the Belarussian-Russian Relationship', *Problems of Post-Communism,* 46:3 (1999), pp. 3-13.

6. Margarita Balmaceda, 'Belarus as a Transit Route: Domestic and Foreign Policy Implications," in Margarita Balmaceda, J. Clem and L. Tarlow, ed., *Independent Belarus: Domestic Determinants, Regional Dynamics and Implications for the West* (Cambridge: HURI/Davis Center for Russian Studies: Harvard University Press, 2002). pp. 162-196.

7. Margarita Balmaceda, Ukraine's Persisting Energy Crisis, *Problems of Post-Communism,* 51:4, (July-September 2004), pp. 40-50.

8. *Belarus as a Gas Transit Country* (Minsk: Research Centre of the Institute of Privatization and Management. German Economic Team in Belarus, 2004).

9. Belarus joins Russia's open access battle, Western Gas Intelligence, 15:2, (January 2004), p.3.

10. Belarus risks cuts in a row over Russian joint venture pipeline, Gas Briefing International, (October, 2002), p.4.

11. Caspian sea and Ukraine's Quest for Energy Autonomy, Geopolitics of Energy, (October 1998).

12. Saul Cohen, *Geography and Politics in a World Divided,* 2nd ed., (Oxford: Oxford University Press, 1973), p. 29.

13. Joseph Crieco, *Cooperation Among Nations,* New York: Cornell University Press, 1990.

14. George Demko and William Wood, 'Introduction: International Relations Trough the Prisms of Geography' in Demko and Wood, eds., *Reordering the World: Geopolitical Perspectives on the Twenty-First Century,* pp. 10-11.

15. Destination clauses remain an issue for accession countries, *European Gas Markets,* 11:06:1, (June, 2004), p.1, 9.

16. James Dougherty and Robert Pfaltzgraff Jr., *Contending Theories of International Relations. A Comprehensive Survey* (London: Longman, 2001).

17. Emerging Threats to Energy Security and Safety, Windsor Intelligence Group/NATO Energy Security Workshop.

18. Javier Estrada et al. *The Development of European Gas Markets. Environmantal, Economic and Political Prospects* (Chichester: Wiley, 1995).

19. EU accession – ten new countries, and a few new problems, *Gas Matters,* (May, 2004).

20. EU Energy Policy, Foreign Policy and Geopolitics: three causes that have not found one another, *Geopolitics of Energy,* 17:8 (August 1995), pp.4-6.

21. EU-Russia energy dialogue making progress, *Gas Matters,* (October, 2002), p.7.

22. EU-Russia summit swiftly followed by green light for North Transgas, *Gas Briefing International,* (November, 2002), pp.1-2.

23. First Yamal pipeline string to hit full capacity by early 2006. *Western Gas Intelligence.*

24. Former Gazprom mangers move towards Itera, *European Gas Markets*, 10:04:1, (April, 2003), p.10.

25. Sherman Garnett and Robert Legvold, eds., *Belarus at the crossroads* (Washington: Carnegie Endowment for Peace, 1999).

26. Gas trader Itera clocks up further sales in the fSU, *European Gas Markets*, 7:04:2, (April, 2000), p.7.

27. Gazprom and POGC consider possibilities of Polish storage, *European Gas Markets*, 11:04:1, (April, 2004), p.13.

28. Gazprom blocks Beregovoye pipeline access to Itera, *European Gas Markets*, 10:05:2, (May, 2003), p.8.

29. Gazprom close to abandoning Yamal 2, its Ukranian bypass, *European Gas Markets*, 9:01:02, (January, 2002), pp.1, 6.

30. Gazprom continues to deny Itera access to Beregovoye, *European Gas Markets*, 11:02:1, (February, 2004), p.11.

31. Gazprom denies that barter payment for gas increasing, *European Gas Markets*, 9:10:2, (October, 2002), p.12.

32. Gazprom drops destination clauses, *European Gas Markets*, 10:10:1, (October, 2003), pp.1,7.

33. Gazprom progresses on pipe schemes, *Western Gas Intelligence*, 13:47, (November 2002), p.3.

34. Gazprom image at risk, *Western Gas Intelligence*, 15:8, (February 2004), pp.1-2.

35. Gazprom reaffirms plan to export 153 bcm to Europe in 2004, *European Gas Markets*, 11:03:1, (March, 2004), p.11.

36. Gazprom rekindles cooperation with Itera, *European Gas Markets*, (October, 2003).

37. Gazprom stops flows to Belarus, *European Gas Markets*, (February, 2004).

38. Gazprom threatens gas producers with halt to pipe access, *European Gas Markets*, 10:10:2, (October, 2003), p.10.

39. Gazprom to make domestic profit in 2004, *European Gas Markets*, 11:03:2, (March, 2004), p.1,6.

40. Moscow wants a direct pipe link to Germany, *Western Gas Intelligence*, 15:4, (January 2004), pp.1-2.

41. Gazprom wants Belarus to pay market price for gas, *European Gas Markets*, 10:10:1, (October, 2003), p.11.

42. The Geopolitics of European Gas, *Geopolitics of Energy*, 1:20, (January 1998), pp.2-4.

43. International consortium to spearhead southward diversion of Yamal-Europe, *European Gas Markets*, 7:10:1, (October, 2000), p.8.

44. The International Energy Agency, *The IEA Natural Gas Security Study* (Paris: OECD, 1995).

45. Itera faces downsizing by Gazprom – again, *Western Gas Intelligence*, 15:20, (May 2004), p.4.

46. Itera offers shares in production company to Gazprom, *European Gas Markets*, 11:05:1, (May, 2004), p.12.

47. Itera suggests share swaps in assets with Gazprom, *European Gas Markets*, (March, 2003).

48. It is all in the pipes: Gazprom's profit lies in transmission, *European Gas Markets*, 7:06:2, (June, 2000), p.8.

49. Igor Ivanov, *New Russian Diplomacy. The decade of the country's foreign policy* (Moscow: OLMA-PRESS, 2002).

50. Taras Kuzio, 'Belarus and Ukraine: democracy building in a grey security zone' in Jan Zielonka and Alex Pravda, ed., *Democratic Consolidation in Eastern Europe. International and Transitional Factors,* 2 vols. (Oxford: Oxford University Press, 2001), vol.2, pp. 455-484.

51. Colin W. Lawson, *Path-dependence and economy of Belarus: the consequences of late reforms,* paper presented at the seminar (Oxford University, Spring 2001).

52. Margot Light, Post-Soviet Russian Foreign Policy: The First Decade in Archie Brown (ed.) *Contemporary Russian Politics: A Reader,* pp. 419-428.

53. The long and the short of Europe's proposed gas security directive, *Gas Matters,* (November 2002), pp.6-11.

54. Alexander Lukashuk, A year on a treadmill, *RFE/RL Research Report,* 2:1 (January, 1993), pp. 64-68.

55. Edward Luttwak, *Endangered American Dream,* New York: Simon and Schuster, 1993, pp. 307-325.

56. Steven J. Main, *Belarus and Russia military cooperation 1991-2002,* Conflict Studies Research Centre, Sandhurst Military Academy, 2002.

57. Makarov's view on prices: 'cheap gas corrupts the consumer'. *Gas Matters,* The interview given by I. Makarov, President of Itera, (November, 2002), pp.12-16.

58. Ustina Markus, Heading off an Energy Disaster, *Transition,* (April, 1995), pp. 10-13.

59. Ustina Markus, Belarus Chooses Dictatorship, *Transition,* (February, 1997).

60. Ustina Markus, Belarus, Ukraine take opposite views, *Transition*, (November, 1996), pp. 20-22.

61. Ustina Markus, 'Still Coming to Terms with Independence', *Transition*, (February, 1995), pp. 47-52.

62. K. McKeough, The natural gas investment climate of the 1990s: more opportunities than capital. *Geopolitics of Energy*, (January, 1994).

63. Kathleen Mihalisko, 'Belarus: retreat to authoritarianism', in Karen Dawisha and Bruce Parrot, ed., *Democratic changes and authoritarian reactions in Russia, Ukraine, Belarus, and Moldova* (Cambridge: Cambridge University Press, 1997).

64. Helen Milner, 'International Theories of Cooperation Among Nations: Strengths and Weaknesses, *World Politics*, (April, 1992).

65. John Mitchell, 'EU Energy Policy, Foreign Policy and Geopolitics: three causes that have not found one another', *Geopolitics of Energy*, (August 1995).

66. The natural gas investment climate of the 1990s: more opportunities than capital, *Geopolitics of Energy*, 16:1, (January 1994), pp.4-7.

67. NorthTransgas pipeline gathers momentum, *European Gas Markets*, 8:01:2, (January, 2001), p.6

68. Mario D. Nuti, 'The Belarus Economy: Suspended Animation between State and Markets', in Stephen White, Elena Korosteleva and John Lowenhardt, ed., *Postcommunist Belarus* (Lanham MD: Rowman & Littlefield, 2004).

69. Mario D. Nuti, 'Making Sense of the Third Way', *Business Strategy Review*, 10:3, (Autumn, 1999), pp. 57–67.

70. Mario D. Nuti, 'Belarus: A Command Economy Without Central Planning', *Russian and East European Finance and Trade*, 26:4, (July–August 2000).

71. Peter Odell, 'The Geopolitics of European Gas', *Geopolitics of Energy*, 26:3 and 26:4, (January, 1998).

72. Bruce Parrot, ed., *Democratic changes and authoritarian reactions in Russia, Ukraine, Belarus, and Moldova* (Cambridge: Cambridge University Press, 1997).

73. POGC threatens to make claim over Gazprom-belarus spat, *European Gas Markets*, 11:03:1, (March, 2004), p.12.

74. Alex Pravda, 'Russia and the 'Near Abroad', in Stephen White, Alex Pravda and Zvi Gitelman, ed., *Developments in Russian Politics 5*, (London: Palgrave, 2001).

75. Russian and Ukrainian gas industries: back to the future, *European Gas Markets*, 8:5:1, (May, 2001), p.8.

76. PGNiG plans to increase supply security following Belarus interruption, *Gas Matters*, (April, 2004), p.17

77. The press-conference given by A. Miller, Gazprom's management committee Chairman, and A. Ryazanov, Deputy Chairman of Gazprom's management committee to Belarussian reporters, 17 May, 2004.

78. The press-conference given by A. Ryazanov, Deputy Chairman of Gazprom's management committee, and A. Medvedev, Deputy Chairman of Gazexport, 2004.

79. Privatization 'the only way to restore' the Ukrainian pipeline, *European Gas Markets*, 11:12:2, (December, 2004).

80. Putin on Gazprom, *Western Gas Intelligence*, 15:11 (March 2004), p.5.

81. Putin stamps his authority on Gazprom, *European Gas Markets*, 8:05:2, (May, 2001), p.1,4.

82. PWC hits dead end in audit of Gazprom-Itera relations, *European Gas Markets*, 8:7:1, (July, 2001), pp.8-9.

83. Ruhrgas to join consortium for Ukraine's gas transit system, *European Gas Markets*, 10:01:1, (January, 2003), p.13.

84. Russia agrees to raise prices as part of EU agreement on WTO accession, *European Gas Markets*, 11:05:2, (May, 2004), p.11.

85. Russia and Belarus agree breakthrough gas deal, *Gas Briefing International*, (June, 2002), pp.2-4.

86. Russia and Belarus agree on joint transit venture on a way to economic union, *Gas Briefing International*, (February, 2003), p.8.

87. Russia and Belarus (and Gazprom and Itera) keep gas flowing – at a price, *Gas Briefing International*, (November, 2002), p.2.

88. 'Russia-Belarus force plans flounder', *Jane's Intelligence Review,* (October 2000), pp. 19-20.

89. Russia cuts transit flows through Belarus – has Belarus become new Ukraine? *Gas Matters*, (February 2004), pp.9-10.

90. Russia natural gas prices may rise 35%, *European Gas Markets*, 8:12:1, (December, 2001), p.12.

91. Russian energy minister proposes CIS an alliance, *European Gas Markets*, 9:10:2, (October, 2002), p.13.

92. Russia's North Europe Gas pipeline moves, *Western Gas Intelligence*, 15:8, (February, 2004), pp.4-5.

93. Russian Gazprom stops flows to Belarus, *European Gas Markets*, 11:02:2, (February, 2004), p.1,6.

94. Russia now threatens a Belarus by-pass, *Gas Matters*, (October 2003), p.11.

95. The Russian Revolution, *European Gas Markets*, 7:04:1, (April, 2000), p.6-7.

96. Russian security of supply not yet under serious threat. *European Gas Markets*, 8:10:1, (October, 2001), p.8.

97. Russia: prices, taxes and the role of Itera, *European Gas Markets*, 8:8:1, (August, 2001), pp.8-9.

98. Russia ups the ante in Yamal re-rerouting row, *European Gas Markets*, 7:08, (August, 2000), p.12.

99. Russia withdraws low gas price concession from Belarus, *Gas Briefing International*, (September 2003), p.5.

100. Richard Sakwa, *Putin: Russia's choice*, London: Routledge, 2002.

101. Richard Sakwa, *Russian Politics and Society* (3rd edition), Routledge (2002), Ch. 16.

102. Richard Sakwa and Mark Webber, 'The Commonwealth of Independent States, 1991-1998: stagnation and survival', *Europe-Asia Studies*, 51:3, 1999, pp. 379-415.

103. Jonathan Stern, *The Future of Russian Gas and* Gazprom, Oxford: Oxford University Press, 2005.

104. Jonathan Stern, *Security of European natural gas supplies. The impact of import dependence and liberalisation,* London: The Royal Institute of International Affairs, Sustainable development programme, July, 2002.

105. Jonathan Stern, 'Soviet and Russian gas: the origins and evolution of Gazprom's export strategy' in Robert Mabro and Ian Wybrew-Bond (eds), *Gas to Europe. The Strategies of the Four Major Suppliers*, Oxford: Oxford University Press, 1999, pp. 135-200.

106. Jonathan Stern, *The Russian Natural Gas `Bubble': Consequences for European Gas Markets,* London: Royal Institute of International Affairs, Brookings, 1995.

107. Michael Stoppard, *A New Order For Gas In Europe*, Oxford: OIES, p. 12.

108. *The Strategy of the relationship between Russian Federation and the European Union in the medium run (1000-2010),* (signed on the 3rd of June, 2000).

109. *Transition Report.* London: EBRD, various years.

110. Paul Schroeder, 'Historical Reality vs. Neo-realist Theory', *International Security,* 19:1, (Summer, 1994), p. 116.

111. Ukraine boosts its domestic gas production, *European Gas Markets,* 8:8:1, (August, 2001), p.12.

112. Ukraine plans to build Black Sea liquefaction plant [$1,2-1,5 bln] – but has given no details on gas sourcing, *Gas Matters,* (April 2004), p.17.

113. Yamal-2 route close to agreement, but SPP will have to find its own finance, *European Gas Markets,* 8:6:1, (June, 2001), p.8

114. И. А. Михайлова-Станюта, 'Открыта ли Белорусская экономика?', *Белорусский банковский вестник,* выпуск 14 (Апрель 2003), стр. 45-49.

115. В. Е. Снапковский и А. В. Шарапо (ред.) *Белорусско-Российские отношения: проблемы и перспективы.* Материалы второго круглого стола белорусских и российских ученых (Минск: БГУ, 2000).

116. Т. Шаклеина, *Белорусия во внешнеолитической стратегии РФ,* Дискуссии о Союзе России и Беларуси в российском политико-академическом сообществе геополитический аспект. Московский общественный научный фонд. Москва, 2000.

117. Margarita Balmaceda et al., eds., *Independent Belarus: Domestic Determinants, Regional Dynamics, and Implications for the West* (Cambridge: Harvard Ukrainian Research Institute and the Davis Centr for Russian Studies, 2003).

Other sources

Белорусский Банковский Вестник, various issues
Эксперт, various issues
*Экономика и бизнес,*various issues
Народная Газета, various issues
Независимая Газета, various issues
Российская Газета, various issues
РИА Новости
The Moscow Times, various issues
Cedigaz, various issues

The Government of the Russian Federation official web site, http://www.gov.ru
The Ministry of Foreign Affairs of the Republic of Belarus, official website
http://www.mfa.gov.by/eng/index.php?d=policy/intpact&id
The National Assembly of the Republic of Belarus, official website, www.house.gov.by.
The President of the Republic of Belarus official web site, http://www.president.gov.by
Gazprom, http://www.Gazprom.ru,
Itera, the independent Russian gas company, http://www.Itera.com
'Novosti Ukrainy' news agency,
http://newsukraina.ru/news.html?nws_id=265773
Interfax news agency, http://interfax.com
European Commission, Documents on the Energy Dialogue, http://www. europa.eu.int?comm/energy_transport/en/lpi_en_3.html
Research Centre of the Institute of Privatization and Management. German Economic Team in Belarus. Minsk, March, 2004. http://www.ipm.by
Radio Free Europe/Radio Liberty (RFE/RL),
IMF, http://www.imf.org.

Appendix -Table 2.1: EBRD Transition Indicators: Belarus and Russia in Comparison

		1995	1996	1997	1998	1999	2000	2001	2002	2003	2004
Index or price liberalisation	Belarus	3.7	3.7	4.0	2.7	2.3	2.3	2.7	2.7	2.7	2.7
	Russia				3.3	3.3	4.0	4.0	4.0	4.0	4.0
Index of FE and trade liberalisation	Belarus	2.0	2.0	1.0	1.0	1.0	1.7	2.0	2.3	2.3	2.3
	Russia				2.3	2.3	2.3	2.7	3.0	3.3	3.3
Index of small scale privatisation	Belarus	2.0	2.0	2.0	2.0	2.0	2.0	2.0	2.0	2.3	2.3
	Russia				4.0	4.0	4.0	4.0	4.0	4.0	4.0
Index of large scale privatisation	Belarus	1.7	1.0	1.0	1.0	1.0	1.0	1.0	1.0	1.0	1.0
	Russia				3.3	3.3	3.3	3.3	3.3	3.3	3.3
Index of enterprise reform	Belarus	1.7	1.7	1.0	1.0	1.0	1.0	1.0	1.0	1.0	1.0
	Russia				2.0	1.7	2.0	2.3	2.3	2.3	2.3
Index of competition policy	Belarus	2.0	2.0	2.0	2.0	2.0	2.0	2.0	2.0	2.0	2.0
	Russia				2.3	2.3	2.3	2.3	2.3	2.3	2.3
Index of infrastructure reform	Belarus	1.0	1.0	1.0	1.0	1.3	1.3	1.3	1.3	1.3	1.3
	Russia				2.0	2.3	2.3	2.3	2.3	2.3	2.7
Index of banking sector reform	Belarus	2.0	1.0	1.0	1.0	1.0	1.0	1.0	1.7	1.7	1.7
	Russia				2.0	1.7	1.7	1.7	2.0	2.0	2.0
Index of reform of non-bank financial institutions	Belarus	2.0	2.0	2.0	2.0	2.0	2.0	2.0	2.0	2.0	2.0
	Russia				1.7	1.7	1.7	1.7	2.3	2.7	2.7

Additional Explanatory Notes after the book went into print

Note 1 on Russian exports to Europe (Table 3.1, p.66):
The total exports figure does not include exports to Baltic countries although they are EU Member States. The total exports figure is somewhat higher than 140.5 bcm cited in other sources because Gazprom includes here spot gas sold in the UK, Belgium and elsewhere. This also explains a higher than otherwise cited figure for Russian exports to Germany (36.1 bcm).

Note 2 on the projected capacity of the North European Gas Pipeline:
Throughout this book we refer to a projected capacity of the NEGP of around 30 bcm, but since the decision has been made to build two strings, the capacity increased to 55 bcm.

Dr. Andreas Umland (Ed.)

SOVIET AND POST-SOVIET POLITICS AND SOCIETY

ISSN 1614-3515

This book series makes available, to the academic community and general public, affordable English-, German- and Russian-language scholarly studies of various *empirical* aspects of the recent history and current affairs of the former Soviet bloc. The series features narrowly focused research on a variety of phenomena in Central and Eastern Europe as well as Central Asia and the Caucasus. It highlights, in particular, so far understudied aspects of late Tsarist, Soviet, and post-Soviet political, social, economic and cultural history from 1905 until today. Topics covered within this focus are, among others, political extremism, the history of ideas, religious affairs, higher education, and human rights protection. In addition, the series covers selected aspects of post-Soviet transitions such as economic crisis, civil society formation, and constitutional reform.

SOVIET AND POST-SOVIET POLITICS AND SOCIETY

Edited by Dr. Andreas Umland

ISSN 1614-3515

1   *Андреас Умланд (ред.)*
Воплощение Европейской конвенции по правам человека в России
Философские, юридические и эмпирические исследования
ISBN 3-89821-387-0

2   *Christian Wipperfürth*
Russland – ein vertrauenswürdiger Partner?
Grundlagen, Hintergründe und Praxis gegenwärtiger russischer Außenpolitik
Mit einem Vorwort von Heinz Timmermann
ISBN 3-89821-401-X

3   *Manja Hussner*
Die Übernahme internationalen Rechts in die russische und deutsche Rechtsordnung
Eine vergleichende Analyse zur Völkerrechtsfreundlichkeit der Verfassungen der Russländischen Föderation
und der Bundesrepublik Deutschland
Mit einem Vorwort von Rainer Arnold
ISBN 3-89821-438-9

4   *Matthew Tejada*
Bulgaria's Democratic Consolidation and the Kozloduy Nuclear Power Plant (KNPP)
The Unattainability of Closure
With a foreword by Richard J. Crampton
ISBN 3-89821-439-7

5   *Марк Григорьевич Меерович*
Квадратные метры, определяющие сознание
Государственная жилищная политика в СССР. 1921 – 1941 гг
ISBN 3-89821-474-5

6   *Andrei P. Tsygankov, Pavel A.Tsygankov (Eds.)*
New Directions in Russian International Studies
ISBN 3-89821-422-2

7   *Марк Григорьевич Меерович*
Как власть народ к труду приучала
Жилище в СССР – средство управления людьми. 1917 – 1941 гг.
С предисловием Елены Осокиной
ISBN 3-89821-495-8

8   *David J. Galbreath*
Nation-Building and Minority Politics in Post-Socialist States
Interests, Influence and Identities in Estonia and Latvia
With a foreword by David J. Smith
ISBN 3-89821-467-2

9   *Алексей Юрьевич Безугольный*
Народы Кавказа в Вооруженных силах СССР в годы Великой Отечественной войны
1941-1945 гг.
С предисловием Николая Бугая
ISBN 3-89821-475-3

10   *Вячеслав Лихачев и Владимир Прибыловский (ред.)*
Русское Национальное Единство, 1990-2000. В 2-х томах
ISBN 3-89821-523-7

11   *Николай Бугай (ред.)*
Народы стран Балтии в условиях сталинизма (1940-е – 1950-е годы)
Документированная история
ISBN 3-89821-525-3

12   *Ingmar Bredies (Hrsg.)*
Zur Anatomie der Orange Revolution in der Ukraine
Wechsel des Elitenregimes oder Triumph des Parlamentarismus?
ISBN 3-89821-524-5

13   *Anastasia V. Mitrofanova*
The Politicization of Russian Orthodoxy
Actors and Ideas
With a foreword by William C. Gay
ISBN 3-89821-481-8

14   *Nathan D. Larson*
Alexander Solzhenitsyn and the Russo-Jewish Question
ISBN 3-89821-483-4

15   *Guido Houben*
Kulturpolitik und Ethnizität
Staatliche Kunstförderung im Russland der neunziger Jahre
Mit einem Vorwort von Gert Weisskirchen
ISBN 3-89821-542-3

16   *Leonid Luks*
Der russische „Sonderweg"?
Aufsätze zur neuesten Geschichte Russlands im europäischen Kontext
ISBN 3-89821-496-6

17   *Евгений Мороз*
История «Мёртвой воды» – от страшной сказки к большой политике
Политическое неоязычество в постсоветской России
ISBN 3-89821-551-2

18   *Александр Верховский и Галина Кожевникова (ред.)*
Этническая и религиозная интолерантность в российских СМИ
Результаты мониторинга 2001-2004 гг.
ISBN 3-89821-569-5

19   *Christian Ganzer*
Sowjetisches Erbe und ukrainische Nation
Das Museum der Geschichte des Zaporoger Kosakentums auf der Insel Chortycja
Mit einem Vorwort von Frank Golczewski
ISBN 3-89821-504-0

20   *Эльза-Баир Гучинова*
     Помнить нельзя забыть
     Антропология депортационной травмы калмыков
     С предисловием Кэролайн Хамфри
     ISBN 3-89821-506-7

21   *Юлия Лидерман*
     Мотивы «проверки» и «испытания» в постсоветской культуре
     Советское прошлое в российском кинематографе 1990-х годов
     С предисловием Евгения Марголита
     ISBN 3-89821-511-3

22   *Tanya Lokshina, Ray Thomas, Mary Mayer (Eds.)*
     The Imposition of a Fake Political Settlement in the Northern Caucasus
     The 2003 Chechen Presidential Election
     ISBN 3-89821-436-2

23   *Timothy McCajor Hall, Rosie Read (Eds.)*
     Changes in the Heart of Europe
     Recent Ethnographies of Czechs, Slovaks, Roma, and Sorbs
     With an afterword by Zdeněk Salzmann
     ISBN 3-89821-606-3

24   *Christian Autengruber*
     Die politischen Parteien in Bulgarien und Rumänien
     Eine vergleichende Analyse seit Beginn der 90er Jahre
     Mit einem Vorwort von Dorothée de Nève
     ISBN 3-89821-476-1

25   *Annette Freyberg-Inan with Radu Cristescu*
     The Ghosts in Our Classrooms, or: John Dewey Meets Ceauşescu
     The Promise and the Failures of Civic Education in Romania
     ISBN 3-89821-416-8

26   *John B. Dunlop*
     The 2002 Dubrovka and 2004 Beslan Hostage Crises
     A Critique of Russian Counter-Terrorism
     With a foreword by Donald N. Jensen
     ISBN 3-89821-608-X

27   *Peter Koller*
     Das touristische Potenzial von Kam''janec'–Podil's'kyj
     Eine fremdenverkehrsgeographische Untersuchung der Zukunftsperspektiven und Maßnahmenplanung zur
     Destinationsentwicklung des „ukrainischen Rothenburg"
     Mit einem Vorwort von Kristiane Klemm
     ISBN 3-89821-640-3

28   *Françoise Daucé, Elisabeth Sieca-Kozlowski (Eds.)*
     Dedovshchina in the Post-Soviet Military
     Hazing of Russian Army Conscripts in a Comparative Perspective
     With a foreword by Dale Herspring
     ISBN 3-89821-616-0

29   *Florian Strasser*
      Zivilgesellschaftliche Einflüsse auf die Orange Revolution
      Die gewaltlose Massenbewegung und die ukrainische Wahlkrise 2004
      Mit einem Vorwort von Egbert Jahn
      ISBN 3-89821-648-9

30   *Rebecca S. Katz*
      The Georgian Regime Crisis of 2003-2004
      A Case Study in Post-Soviet Media Representation of Politics, Crime and Corruption
      ISBN 3-89821-413-3

31   *Vladimir Kantor*
      Willkür oder Freiheit
      Beiträge zur russischen Geschichtsphilosophie
      Ediert von Dagmar Herrmann sowie mit einem Vorwort versehen von Leonid Luks
      ISBN 3-89821-589-X

32   *Laura A. Victoir*
      The Russian Land Estate Today
      A Case Study of Cultural Politics in Post-Soviet Russia
      With a foreword by Priscilla Roosevelt
      ISBN 3-89821-426-5

33   *Ivan Katchanovski*
      Cleft Countries
      Regional Political Divisions and Cultures in Post-Soviet Ukraine and Moldova
      With a foreword by Francis Fukuyama
      ISBN 3-89821-558-X

34   *Florian Mühlfried*
      Postsowjetische Feiern
      Das Georgische Bankett im Wandel
      Mit einem Vorwort von Kevin Tuite
      ISBN 3-89821-601-2

35   *Roger Griffin, Werner Loh, Andreas Umland (Eds.)*
      Fascism Past and Present, West and East
      An International Debate on Concepts and Cases in the Comparative Study of the Extreme Right
      With an afterword by Walter Laqueur
      ISBN 3-89821-674-8

36   *Sebastian Schlegel*
      Der „Weiße Archipel"
      Sowjetische Atomstädte 1945-1991
      Mit einem Geleitwort von Thomas Bohn
      ISBN 3-89821-679-9

37   *Vyacheslav Likhachev*
      Political Anti-Semitism in Post-Soviet Russia
      Actors and Ideas in 1991-2003
      Edited and translated from Russian by Eugene Veklerov
      ISBN 3-89821-529-6

38   *Josette Baer (Ed.)*
      Preparing Liberty in Central Europe
      Political Texts from the Spring of Nations 1848 to the Spring of Prague 1968
      With a foreword by Zdeněk V. David
      ISBN 3-89821-546-6

39   *Михаил Лукьянов*
      Российский консерватизм и реформа, 1907-1914
      С предисловием Марка Д. Стейнберга
      ISBN 3-89821-503-2

40   *Nicola Melloni*
      Market Without Economy
      The 1998 Russian Financial Crisis
      With a foreword by Eiji Furukawa
      ISBN 3-89821-407-9

41   *Dmitrij Chmelnizki*
      Die Architektur Stalins
      Bd. 1: Studien zu Ideologie und Stil;
      Bd. 2: Bilddokumentation
      Mit einem Vorwort von Bruno Flierl
      ISBN 3-89821-515-6

42   *Katja Yafimava*
      Post-Soviet Russian-Belarussian Relationships
      The Role of Gas Transit Pipelines
      With a foreword by Jonathan P. Stern
      ISBN 3-89821-655-1

FORTHCOMING (MANUSCRIPT WORKING TITLES)

Stephanie Solowyda
Biography of Semen Frank
ISBN 3-89821-457-5

Margaret Dikovitskaya
Arguing with the Photographs
Russian Imperial Colonial Attitudes in Visual Culture
ISBN 3-89821-462-1

Stefan Ihrig
Welche Nation in welcher Geschichte?
Eigen- und Fremdbilder der nationalen Diskurse in der Historiographie und den Geschichtsbüchern in der Republik
Moldova, 1991-2003
ISBN 3-89821-466-4

Sergei M. Plekhanov
Russian Nationalism in the Age of Globalization
ISBN 3-89821-484-2

Robert Pyrah
Cultural Memory and Identity
Literature, Criticism and the Theatre in Lviv - Lwow - Lemberg, 1918-1939 and in post-Soviet Ukraine
ISBN 3-89821-505-9

Andrei Rogatchevski
The National-Bolshevik Party
ISBN 3-89821-532-6

Zenon Victor Wasyliw
Soviet Culture in the Ukrainian Village
The Transformation of Everyday Life and Values, 1921-1928
ISBN 3-89821-536-9

Nele Sass
Das gegenkulturelle Milieu im postsowjetischen Russland
ISBN 3-89821-543-1

Julie Elkner
Maternalism versus Militarism
The Russian Soldiers' Mothers Committee
ISBN 3-89821-575-X

Maryna Romanets
Displaced Subjects, Anamorphosic Texts, Reconfigured Visions
Improvised Traditions in Contemporary Ukrainian and Irish Literature
ISBN 3-89821-576-8

Alexandra Kamarowsky
Russia's Post-crisis Growth
ISBN 3-89821-580-6

Martin Friessnegg
Das Problem der Medienfreiheit in Russland seit dem Ende der Sowjetunion
ISBN 3-89821-588-1

Nikolaj Nikiforowitsch Borobow
Führende Persönlichkeiten in Russland vom 12. bis 20 Jhd.: Ein Lexikon
Aus dem Russischen übersetzt und herausgegeben von Eberhard Schneider
ISBN 3-89821-638-1

Anton Burkov
The Impact of the European Convention for the Protection of Human Rights and Fundamental Freedoms on Russian Law
ISBN 3-89821-639-X

Martin Malek, Anna Schor-Tschudnowskaja
Tschetschenien und die Gleichgültigkeit Europas
Russlands Kriege und die Agonie der Idee der Menschenrechte
ISBN 3-89821-676-4

Claudia Dathe, Anastasija Grynenko und Andreas Umland (Hrsg.)
Die Terminologie des ukrainischen und deutschen Gerichtswesens im Vergleich
Eine übersetzungswissenschaftliche Analyse
ISBN 3-89821-691-8

Christopher Ford
Borotbism: A Chapter in the History of Ukrainian Communism
ISBN 3-89821-697-7

Taras Kuzio, Paul D'Anieri (Hrsg.)
Aspects of the Orange Revolution I: Regime Politics and Democratization in Ukraine
ISBN 3-89821-698-5

Bohdan Harasymiw, Oleh S. Ilnytzkyj (Hrsg.)
Aspects of the Orange Revolution II: Analyses of the 2004 Ukrainian Presidential Elections
ISBN 3-89821-699-3

Togzhan Kassenova
Cooperative Security in the Post-Cold War International System
The Cooperative Threat Reduction Process
ISBN 3-89821-707-8

Andreas Langenohl
Political Culture and Criticism of Society
Intellectual Articulations in Post-Soviet Russia
ISBN 3-89821-709-4

Marlies Bilz
Tatarstan in der Transformation, 1988-1994
ISBN 3-89821-722-1

Thomas Borén
Meeting Places in Transformation
ISBN 3-89821-739-6

Lars Löckner
Sowjetrussland in der Beurteilung der Emigrantenzeitung 'Rul', 1920-1924
ISBN 3-89821-741-8

Ekaterina Taratuta
The Red Line of Construction
Semantics and Mythology of a Siberian Heliopolis
ISBN 3-89821-742-6

Stina Torjesen, Indra Overland (Hrsg.)
International Election Observers in Azerbaijan
Geopolitical Pawns or Agents of Change?
ISBN 3-89821-743-4

Series Subscription

Please enter my subscription to the series *Soviet and Post-Soviet Politics and Society*, ISSN 1614-3515, as follows:

❐ complete series   OR   ❐ English-language titles
                                                     ❐ German-language titles
                                                     ❐ Russian-language titles

starting with
❐ volume # 1
❐ volume # ___
     ❐ please also include the following volumes: #___, ___, ___, ___, ___, ___, ___
❐ the next volume being published
     ❐ please also include the following volumes: #___, ___, ___, ___, ___, ___, ___

❐ 1 copy per volume   OR   ❐ ___ copies per volume

Subscription within Germany:

You will receive every volume at 1st publication at the regular bookseller's price – incl. s & h and VAT.
Payment:
❐ Please bill me for every volume.
❐ Lastschriftverfahren: Ich/wir ermächtige(n) Sie hiermit widerruflich, den Rechnungsbetrag je Band von meinem/unserem folgendem Konto einzuziehen.

Kontoinhaber: _____ Kreditinstitut: _____
Kontonummer: _____ Bankleitzahl: _____

International Subscription:

Payment (incl. s & h and VAT) in advance for
❐ 10 volumes/copies (€ 319.80)   ❐ 20 volumes/copies (€ 599.80)
❐ 40 volumes/copies (€ 1,099.80)
Please send my books to:

NAME_____ DEPARTMENT_____
ADDRESS _____
POST/ZIP CODE_____ COUNTRY _____
TELEPHONE _____ EMAIL_____

date/signature_____

A hint for librarians in the former Soviet Union: Your academic library might be eligible to receive free-of-cost scholarly literature from Germany via the German Research Foundation. For Russian-language information on this program, see
     http://www.dfg.de/forschungsfoerderung/formulare/download/12_54.pdf.

Please fax to: **0511 / 262 2201 (+49 511 262 2201)**
or mail to: *ibidem*-Verlag, Julius-Leber-Weg 11, D-30457 Hannover,Germany
or send an e-mail: ibidem@ibidem-verlag.de

ibidem-Verlag

Melchiorstr. 15

D-70439 Stuttgart

info@ibidem-verlag.de

www.ibidem-verlag.de
www.edition-noema.de
www.autorenbetreuung.de

www.ingramcontent.com/pod-product-compliance
Lightning Source LLC
Chambersburg PA
CBHW050531270326
41926CB00015B/3170